WHERE DO I PUT THE DECIMAL POINT?

WHERE DO I PUT THE DECIMAL POINT?

How to Conquer Math Anxiety and Let Numbers Work For You

**ELISABETH RUEDY
and SUE NIRENBERG**

AVON BOOKS ◆ NEW YORK

Grateful acknowledgment is made for the right to reprint "The Need to Win," translated by Thomas Merton, in *The Way of Chuang Tzu*. Copyright © 1965 by the Abbey of Gethsemani. Reprinted by permission of New Directions Publishing Corporation.

AVON BOOKS
A division of
The Hearst Corporation
1350 Avenue of the Americas
New York, New York 10019

The Henry Holt and Company edition contains the following Library of Congress Cataloging in Publication Data:

Ruedy, Elisabeth.
Where do I put the decimal point? : how to conquer math anxiety and increase your facility with numbers / Elisabeth Ruedy and Sue Nirenberg. — 1st ed.
    p.   cm.
Includes bibliographical references.
1. Mathematics—Study and teaching—Psychological aspects. 2. Mathematics. I. Nirenberg, Sue. II. Title.
QA11.R743   1990
510′.71—dc20
                                                                89-48728
                                                                CIP

First Avon Books Trade Printing: March 1992

AVON TRADEMARK REG. U.S. PAT. OFF. AND IN OTHER COUNTRIES, MARCA REGISTRADA, HECHO EN U.S.A.

Printed in the U.S.A.

OPM   10 9 8 7 6 5 4 3 2 1

CONTENTS

ACKNOWLEDGMENTS

Writing this book was a complex process, incorporating years of work and experience, moments of insight and conversations with people who inspired me.

My full appreciation goes to my husband and children, who lived through the "summer of the book" with good humor and who gave me all the essentials: love, support, and time. My daughter Nicole was involved in every phase. She proved a capable reader-editor, and her sense of humor made the dry stretches bearable and fun. My son Lucas was an unstinting friend and support. My husband read numerous drafts and cooked many dinners.

As every teacher knows, I owe much gratitude to my students. Many of them have become my friends and all of them have taught me. Special thanks to Matthew Jones for his support in the early phases of this project and for helping me clarify my point of view.

Sue Nirenberg and I thank our editor, Cynthia Vartan, who was insightful and generous at all times, and our agent, Ivy Stone, who could always be relied on for excellent advice and friendship.

Along my path there have been many other people who encouraged and befriended me, and whose life and work inspired me. I can only mention a few: J. Krishnamurti, George Pólya, Earlene Spellman, Marilyn Russakoff, George Murray, Jay, Malik, and Malcolm Murray, Joe Silberfarb, Lili Toborg, Coco Van Meerendonk, Stanley Bosworth, Peter Lynn, Theseus Roche, Margaret Jones, Ruth Van Doren, Mavis Sutton, Susan Xenarios and all my friends at RIP, Letitia Chamberlain, Elaine Sheffler and Florence Tannen, and all the people at Party Cake and The Hungarian Pastry Shop. My thanks to all.

Elisabeth Ruedy

WHERE DO I PUT THE DECIMAL POINT?

. . INTRODUCTION

I grew up seeing math as a simple, down-to-earth endeavor. From an early age, I loved numbers and was fascinated by the way things fell into place in nature and in math. What is more unusual is that I was just as interested in people around me who could *not* do math. Why would perfectly bright friends of mine have trouble adding and subtracting? My conviction today, that there is no reason for anybody who can function in his or her life to feel afraid of math, began way back then. My compassion for math-anxious friends was helped along by my second-grade teacher, who pulled girls several inches off the floor by their hair when they could not answer a math problem. That both scared and angered me. Determined never to have to live through a hair-pulling scene again, I invited all interested classmates to meet me at the garden fence after school. I drew pictures for them. I counted on my fingers and on theirs. I explained how to make it all easier until a couple of months later I had won my battle. The hair-pulling was over.

This first math-anxiety clinic still holds meaning for me. My methods and views have become more informed and sophisticated, but the basic spirit of my work is the same. I abhor suffering. I will work hard to spare people humiliation and pain. I resent it that so many people are made to feel stupid around numbers—girls in particular. I believe that everybody can learn to do math to the degree necessary to use it in their lives. People's inclinations and talents go nicely together. If math is not your thing, you're not likely to aspire to professions riddled with math. Yet you certainly can pick up the basics to function as a competent worker or citizen. Having grown up in Switzerland and having been taught by some of the best math teachers, I am eager to share the ideas and methods that made math easy for me to learn and teach, particularly since so few of them are currently used in this country.

I love teaching math by imagining, drawing, thinking in context, trial and error. In short, I love teaching it away from the dry, left-brain, step-by-step, there-is-only-one-way methodology that scares away the more creative, dreamy child or adult not so interested in precision and linear thinking (but bent on fully understanding the task at hand). The material presented in this book is similar to that presented in my workshops on math anxiety and on learning theory.

I have studied learning theory, recent brain research, problem-solving methods, and even bodywork—Alexander and Feldenkrais—all just to get closer to understanding why teachers, parents, schools, and textbooks abandon students to math anxiety and/or math incompetence.

Let me hasten to add that the two don't necessarily go together. I have met highly competent math students who were afraid of every mathematical step they took, and rather math-illiterate adults who were cheerfully unconcerned about their math weakness. In my classes and workshops for both children and adults, I have proved again and again that everyone who can find his or her way to class can also learn basic math. If you think you are math-anxious, this book will help you to zero in on your blind spots, give you the proper confidence-building techniques, and teach you math in a new, easy-to-understand way. Instead of having you memorizing formulas and procedures, I will teach you to use your visual skills and common sense—yes, you do have them!—to learn to analyze a problem until the solution falls into your lap. This will raise your level of problem-solving skills rather than just teach you immutable formulas. In this book I'll prove to you that women *can* do as well in math as men, that math requires intuition as well as logic, and that there's more than one way to do a math problem. I'll also show you that it's not always important to get the answer exactly right, that you may count on your fingers, and that not all mathematicians do problems quickly in their heads. I'll distinguish between math and plain numerical information, which is often confused with math, as are maps, charts, and computer programs.

Math anxiety has many sources but usually starts in childhood. It is therefore important for anyone who has it to remember what happened to him or her as a child. I have never found a little child who did not love numbers. Numbers instill a sense of power in children, I think, and continue to do so in adults, too. In all my years as a classroom teacher, seminar leader, and math tutor, I have never lost confidence in the ability of my students. It has been most rewarding work. It is music to my ears when I hear the sighs of relief, the *ohs* and *ahs* at easy ways of doing things, and amazed exclamations like "You mean that's all there is?" or "You mean I can just imagine it, think about it, and do it my way?"

Where Do I Put the Decimal Point? (the question that math-anxious people most persistently ask) is a program for overcoming math anxiety. Part I tells you how to identify and access your own math anxiety through a quiz and exercises; Part II describes the math myths and math blocks that stand between you and competence in math; Part III gives you Confidence-Building Techniques to help you cope with your anxiety; in Part IV you'll relearn essential math skills you may have forgotten, or learn them if you've never learned them before; in Part V you'll see how these basic strategies will serve you when math comes up in everyday life; and Part VI ("Afterthoughts") describes the happy consequences of what you've learned in this book.

I sincerely hope that *Where Do I Put the Decimal Point?* will guide you to some moments of relief, cheerfulness, confidence, and competence, and my only regret is that I won't be there to witness your emergence from an anxious, avoiding view of math to the sense of fun and beauty that math offers. I hope you will lose your fear of numbers and their authoritative "halo," and that you will begin to see math for the simple, down-to-earth endeavor it is.

I...
PINPOINTING
THE PROBLEM

1.. THE PROBLEM

If we had to teach our children to walk and talk, they would never learn. Wise people inside and outside the field of education have kept their eyes on that statement to keep themselves humble and open to new ideas. In this country we start teaching math at earlier and earlier ages, and we do well by some students but badly by a majority of them. Many leave school undertaught and badly math-anxious. Math anxiety, which is widespread and poses a serious problem in our society, often stems from early childhood experiences, including intimidation and humiliation. It is reinforced by cultural and family messages and transmitted by teachers who suffer from it themselves. More frequent in women than men, it is manifest in self-sabotaging behavior that leads to unsatisfactory work lives. In this book I'll show you that math anxiety can be modified and even reversed.

In the beginning of 1989, a spate of media material appeared pointing to the American deficiency in mathematics. The National Research Council, an arm of the National Academy of Sciences and of the National Academy of Engineering, reported that most American students leave school lacking the mathematical skills to meet job demands or to continue their education. And the Education Testing Service found that American teenagers scored lowest in mathematics in an international survey. Within a few months Phil Donahue aired a show about Asian students' superiority in math and science, and Barbara

Walters narrated two documentaries on "Why America Flunks" and "Teaching Kids to Think." The Mathematical Sciences Education Board published *Everyone Counts: A Report to the Nation on the Future of Mathematics Education,* which reflected the thinking of seventy leading Americans, among them classroom teachers, college and university faculty and administrators, research mathematicians and statisticians, all of whom were prepared to participate actively in the long-term work of rebuilding mathematics education in the United States. People no longer have the luxury of ignoring math, because, as the report states:

Mathematics is the key to opportunity. No longer just the language of science, mathematics now contributes in direct and fundamental ways to business, finance, health, and defense. For students, it opens doors to careers. For citizens, it enables informed decisions. For nations, it provides knowledge to compete in a technological economy. To participate fully in the world of the future, America must tap the power of mathematics.

About this same time, a book titled *Innumeracy: Mathematical Illiteracy and Its Consequences,* by John Allen Paulos, was published. Dr. Paulos, a professor of mathematics at Temple University, defines "innumeracy" as the inability to understand the basic elements of mathematics (numbers, logic, and probability). He wrote *Innumeracy* to challenge the literate layperson who feels inadequate in situations that require some facility and understanding of math. "Few educated people will admit to being completely unacquainted with the names Shakespeare, Dante, or Goethe," Dr. Paulos says, "yet most will openly confess their ignorance of Gauss, Euler, or Laplace, in some sense their mathematical analogues. . . . I'm distressed by a society which depends so completely on mathematics and science and yet seems so indifferent to the innumeracy and scientific illiteracy of so many of its citizens.'' And he goes on to ask, ''Why is innumeracy so widespread even among otherwise educated people? The reasons, to be a little simplistic, are poor education, psychological blocks, and romantic misconceptions about the nature of mathematics.''

Astonishingly (since the general reading public does not usually seize upon books about math), his book hit the *New York Times* best-seller list, where it remained for more than twenty weeks.

The abovementioned works, however, comment mostly on math *deficiency*. Intimately connected with and underlying math deficiency is math *anxiety*. On the surface we don't see much math *anxiety*. We see massive math deficiency, a phenomenon similar to illiteracy. Deeply ashamed of not being able to handle numbers, people stop learning early. They develop a level of math deficiency that can be as damaging to themselves (and ultimately to society) as is illiteracy.

Because widespread shame and anxiety about not knowing math prevents people from learning it, a few good courses won't remedy the problem. The fact is that people hide their incompetence or sit through remedial courses (doing as badly as they did the first time around), afraid to appear stupid by asking the questions that trouble them, such as:

Where do I put the decimal point?
What is a percent? Is it a number or money or what?
What does *prorate* mean?
I never could learn my tables. Do you think I can learn any math?

Women and minorities suffer from math anxiety more often than men because of low external motivation. Math is considered a male ''insider'' subject, technical and connected to money and power. Many women, in talking about their mathematical pasts, recall their fathers trying to teach them math. Other women, describing their current difficulties with math, will, like Lisa, age thirty, the advertising director of a popular women's magazine, talk about relying on boyfriends or husbands:

> My husband is a math whiz. He can do any math you ask him about. Even the simplest math to him looks like a maze of numbers to me. I always look for the easiest way. To me, it's the most difficult subject ever to learn.

And Vivian, age forty, an operating-room nurse, once considered moving in with a man she did not like much, just to have somebody around to take care of the finances.

> In my personal life, I have always delegated financial responsibility to my current "significant other." I was actually ready to live with Pete, a man I'd been dating but didn't particularly like. I thought I had to have someone to make financial decisions and deal with the checkbook. Then I decided that was crazy. We didn't have much in common, and a few weeks ago I told him I thought we should go our separate ways. Now there's no significant other, and I'll deal with money myself. How could I think of moving into a relationship with finances as my chief motivator? I want to take care of myself, and part of that is taking care of myself financially. So, I'm biting the bullet. I've bought a book called *How to be a Financially Secure Woman*, and I've also been looking around for a good math workbook.

Because of their low visibility in the classroom, women and minorities receive less than adequate attention, low expectations, and meager encouragement. As adults, thousands of competent women can't balance a checkbook, figure the tip in a restaurant, or deal in a mathematical way with discounts or income taxes. They are hesitant and ineffective in negotiating salary raises, or in interpreting graphs and computer printouts. The worst toll that math anxiety takes is that it forces many women to make negative career choices far away from math and therefore from their favorite fields as well. Joanna, forty-five, a sixth-grade teacher, says:

> I ended up as an elementary school teacher because I could not qualify for any of the fields I liked. I would have loved to go into psychology or science, but of course it was impossible because of math anxiety. As the ultimate irony, I had to teach math to my sixth graders, and I was again terrified.

Many times a budding medical student switches to a "softer" field like sociology, or a psychologist switches to elementary school teaching. Marge, a fifty-six-year-old executive secretary to the president of a large printing company, believes that persistent math anxiety crippled her intellectual performance during her adult years. Admitting she made a negative career choice, Marge says, "I wanted to get a degree in nutrition, but I decided against it because the degree program required chemistry."

In a search for prestige and status, many disciplines have upgraded their math requirements, driving away women and minorities who are far more comfortable with qualitative than quantitative information. This culture's overvaluation of numbers has scared even more people away from math. People's fear of mathematics, and their awe of those who speak in numbers, makes them victims of manipulation by statistics. Many people will trust any numerical information without questioning how the numbers were arrived at. A healthy dose of irreverence goes a long way toward freeing you from the tyranny of manipulation by numbers and data. It also does wonders in helping you get rid of math anxiety.

One of the first tasks in overcoming your math anxiety is to strip mathematics of some of its undeserved "halos." That math is given an exalted position in many people's minds is shown in statements such as "Mathematics is the queen of science," and "Mathematics is the ultimate truth." When pressed on this point, most people find it difficult to define what is meant by the word *mathematics*.

WHAT IS MATH?

In Western culture, many people (the math-anxious ones) have stretched the term *mathematics* far beyond its justified limits, to include all manner of sciences, technologies, and business procedures. In fact, all those technical gadgets with pushbuttons and flashing lights have little to do with math. Today's technology, which may frighten you, is not really math, but rather a vast field encompassing everything from an artificial lung to a typewriter to a radio to any state-of-the-art machinery. Once you realize that technology and many other things do not require math knowledge as much as a good instruction manual, you have begun to get a handle on this "inappropriate" math anxiety.

Well, what *is* math?

In my workshops I usually start the first session by asking for a definition. Everyone fidgets, but a brave few confront the issue.

"Math is an attempt to regulate things," volunteers one student.

"Math is when there's one right answer," another calls out.

"Math is a language of measurement," suggests yet another.

"Math is something to stay away from," says the class wit.

Finally someone says, "Math is a conceptual framework that can be used as a tool for analysis and discovery."

BINGO!

Math is a language to describe patterns and structures, numerical and geometrical relationships. Unlike spoken language, which leaves plenty of room for misunderstanding, math is a totally unambiguous language that helps people state clearly what they mean. For some people, math is a perfect escape, just because things are so clear and precise, and usually there is an answer to every question.

Math is one of the oldest human endeavors. We have proof that every civilization we know of has used math inherited from other cultures, and added new insights of its own. The Greeks became heirs to the vast mathematical knowledge accumulated by the Babylonians and the Egyptians. The Europeans, during the Renaissance, seized on all mathematical and scientific lore from antiquity.

Math is alive and expanding even in these days. New ideas and whole new branches of inquiry are continually being added to keep up with new mathematical questions and with the progress in physics and chemistry.

Competence in math can change your life. Learning math gives you a new angle on looking at architecture, music, plants, and sculpture. With new confidence and a new way of looking at life, your self-image improves. When math holds so much promise, why, then, is it so often accompanied by anxiety?

2.. WHAT IS MATH ANXIETY ANYWAY?

Math anxiety, a student once told me, is a large, forbidding, grotesque monster that follows her through life. It slams her over the head, grabs her shoulders, chokes her so that she gasps for breath. As a result of the Math Monster's abusive treatment, she cannot think. Math-anxious people shut down and fail to listen or function. Akin to social anxiety and performance anxiety, with their physical symptoms of pounding heartbeat, racing pulse, and churning stomach, math anxiety evokes continual questioning: *Can I do it? What will they think of me? What will I do if I flop?*

Does this sound familiar?

Does it sound like *you*?

How often does the terrible Math Monster attack? When did the creature enter your life? Math anxiety is not genetic; no one is born with it, and no one will die from it. But along life's way, crucial scenes, powerful and hard to diffuse, may cause derailments. These scenes go very deep and stay with you forever, unless you acknowledge and deal with them. Carolyn, thirty-eight, a lab technician in a pediatrician's office, remembers well:

I HATE MATH. That's all there is to it. Ever since first grade I have had a miserable time with it. One of the most important subjects in school, and I could never pass it. That's so frustrating, and when something is frustrating you try to avoid it. Well, I've done my best in trying to stay away from math. But one way or another, we bump into each other. We've just never had a very good relationship, math and I. We don't understand each other.

Maybe it was my teachers. Old, crumpled nuns are not the best of teachers for young children, in my opinion. They have the most marvelous way of answering every question with, "Sit down and be quiet, you're disturbing the class." Disturbing the class! I thought that's what teachers were for—to answer your questions. Especially math teachers. I mean, math involves more questions than answers. But that was a long time ago.

Madelyn, forty, an associate registrar of a large university, remembers:

My problem has been around as long as I can remember. I recall my father as a math wizard. While we were having breakfast in the kitchen, he would drill me. I graduated to junior high school and my mother hired a tutor. Now I need a career change, and math stands in my way. Math and stammering have managed to choke me to death. Help!

INTIMIDATION BY WEALTH

Sometimes math anxiety starts later in life. Maybe you are intimidated by powerful real-life or media heroes or heroines. Donald Trump, Henry Kravitz, and Rupert Murdoch, for example, are the very embodiment of power in America today (maybe we'll have some nobler heroes in the 1990s). Math-anxious people assume these moguls make and spend large sums of money without referring to notes or calculators. After all, they appear to manipulate situations by spewing out numbers without a moment's hesitation. Remember Joan Collins as the wicked and powerful Alexis in television's "Dynasty"? Did we see her look anything up or write anything down? Never.

Many people in real life are impressed by people who figure easily in their heads. For example, William, twenty-five, a department store manager, is impressed by his friend Charles's ability in math. And Charles, in turn, is dismayed by William's inability to calculate in his head. According to William:

I can't do head calculations, except for the most elementary. Recently a friend of mine, who rattles off calculations in his head involving percentages in a variety of applications, remarked to me when I couldn't follow him, "You do not function in math."

INDIFFERENCE TO MATH

Perhaps you'd rather drop the whole issue of math. Frankly, math is not your thing and you couldn't care less. "Leave me alone" you might be saying. You don't give a hoot about competence.

Robert, a humanities instructor in college, says:

I did well in English and was relegated to that group of people who did well in the humanities. I just didn't have a mathematically adept mind, but neither did other liberal-arts types. In college I applied for exemptions from math and science courses on the grounds that I'd had a few short stories published and wanted to pursue an area amenable to my skills. I've published an article in a travel magazine and have even written the script for a Hollywood movie. I mention these points to bolster my claim to having a well-massaged right cerebral cortex, but as for the left cortex, I have staggered through life as a mathematical illiterate.

You *can* exist in this culture without knowing much math. My friend Sandra, for example, a successful writer whose accountant handles her money, is totally indifferent to numbers. While I am totaling the check when we meet for lunch, Sandra tips by dropping dollar after dollar onto the table. I once questioned her extravagant tipping. Matter-of-factly and without a trace of embarrassment, Sandra outlined her tipping policy: "I leave a dollar for coffee or several dollars for a meal, whatever the total. I don't want to bother." Accepting whatever change she is offered without counting, Sandra always overtips cab drivers, who naturally appreciate her "generosity." The loss of something like ten dollars a day is of no concern to her. (If you're poor, Sandra's method is not for you. The rich seem to have a lot of breaks.)

Sandra reminds me of the college professor who wrote to *People* magazine about the story on Dr. Paulos's *Innumeracy.* Disagreeing with the "eminent math professor" (John Paulos) about the difficulties of "innumeracy," the professor admits that he uses his fingers when computing bank deposits, and is ignorant about percentages, long division, algebra, and geometry. He simply overcomes his mathematical illiteracy in restaurants by asking dinner guests to figure the restaurant bills, while an accountant figures his income taxes, "which are considerable, despite my 'innumeracy.' "

Maybe, like the contented professor, you're a professional whose math competence is limited, but unlike him, you're less than cheerful about your mathematical illiteracy. For a while, everything is dandy. No one knows. But when you get a promotion, trouble brews. Funds, budgets, graphs, and tables intrude into your life. Advancement has made your math deficiency surface, and your new anxiety becomes a major handicap. You worry all day long that you'll be exposed as a fraud. You're suffering from the well-known "Impostor Syndrome." Ann, thirty-four, an assistant in marketing at a large advertising company, could handle her employer's personal bank account while she was a secretary, but actually turned down a position (and a promotion) in the business manager's office. "It involved working exclusively with figures," Ann said.

I meet men and women like Ann all the time in my workshops. They achieve responsible positions with their limited knowledge of math, but suddenly that increased responsibility causes them terror.

Bonnie, thirty-seven, an editor in a prestigious publishing house, was terrified of budgets. Bumped up three levels in management and now a division head, Bonnie was scared in management meetings where the department heads would fight over money, personnel, and allocations. "They could just sling out percentages without any effort at all," said Bonnie.

Dolores, forty-three, a Bolivian in Chicago, is a major lending officer for a bank. Dolores panicked when she had to figure interest on foreign loans.

Leo, forty-four, a high-level trader in a major financial house, was unable to key in certain percentages into the calculator, so he taped to his desktop an index card listing standard percentages.

VANESSA'S STORY

Then there was Vanessa, who came to me for help because she desperately wanted to go to a graduate school with a requirement in statistics and needed a high score on the Graduate Record Exam. Vanessa, twenty-three, the oldest of three daughters, was a brilliant young woman. But by constantly calling her "stupid," particularly when it came to numbers, Vanessa's father had instilled in her a fear of math. Although many math-anxious people remember exactly when their anxiety started—coming back to school after a long absence, encountering a mean teacher, moving to a new town—Vanessa *could not remember a time when she felt comfortable with numbers*. It had taken her years to learn to tell time. (Most children acquire this skill in the first or second grade.) Nevertheless, Vanessa, having skillfully avoided math throughout school and college, graduated with high academic honors. However, whether math anxiety starts with a bang or a whimper, math incompetence is cumulative. The gap in knowledge grows steadily as time goes on, and Vanessa had lost just about all her time.

Since graduation, Vanessa had worked as a mental-health aide at a Philadelphia hospital. To go further in the work she loved, she needed to meet certain academic requirements to qualify as a clinical psychologist. Four previous tutors had thrown their hands up in despair. Vanessa's goals seemed too ambitious; her math ability was at third-grade level. (You'll meet Vanessa again on page 77, and learn how, with iron determination, she did finally exorcise the Math Monster.)

SELECTIVE MATH ANXIETY

While Vanessa's math anxiety was constant, some people have "selective" math anxiety, anxiety that engulfs them only under special mathematical circumstances. Are you one of these people? Sometimes you *can* do math and sometimes you *can't;* sometimes you *won't* do math and sometimes you *will*. Math anxiety comes and goes in your life.

ANYTHING BUT FRACTIONS!

You may be highly ambitious about math, and an excellent conceptual thinker, like Patrice, twenty-two, a travel agent, who is quite sophisticated algebraically and can anticipate my next move in *higher* math. She sees where we're going; parabolas, ellipses, and all manner of interesting functions delight her. Ah, but there's a snag. The moment Patrice spots a fraction by itself (or in an equation), she panics. She pales, goes crazy, tunes out, *can't* do it, *won't* do it.

My main concern is that I do not remember anything about fractions or decimals. Sometimes I have to pay bills from foreign countries converting from English pounds and French francs into dollars, and I get confused with the decimal point.

And Dominique, thirty-four, the manager of a highly specialized antique furniture gallery, says:

I'm terrible with percents and fractions. As the manager of the gallery, I feel nervous when I have to check my answers two or three times while customers wait. Usually they strut back and forth impatiently outside my office, or stand over me, obviously irritated by my slowness.

Loretta, forty-two, a budding designer, found that the Math Monster appeared to her in fashion design school:

I had to know math periodically. It was murder for me. When I see numbers I go crazy, and my teachers were very strict about things like figuring yardages and widths.

STREET-SMART MATH COMPETENCE

Maybe you can't do school math, but if money or concrete objects are involved, your math is just fine, thanks. Ernie, a middle-aged minister who had grown up with little formal education, had street smarts. Like many street-smart people, he could do any problem concerning money with amazing speed and accuracy. He understood percentages intuitively, and amazed all the other students in the class by using commonsense methods rather than formulas to solve problems. Totally self-taught, he had escaped all the routine, the abstractions, and the indoctrination that make math hard for others.

ACADEMIC MATH COMPETENCE

Maybe you can do sophisticated classroom algebra, but the combination of real-life math and money is threatening. You cannot keep your checkbook up to date, think about investments, look at *The Wall Street Journal,* or deal with taxes. If you are a woman, possibly your husband or father takes care of all your mathematical chores. Women often associate money with power and control (reasonably so!), and since they are reluctant to take charge, they avoid money matters. Sylvia, forty-four, executive secretary to the marketing director of a toy manufacturer, says:

> I have a very simple income tax, and I'm ashamed to say that when my accountant got sick one year, I asked my boyfriend to figure it out. I positively cannot change the proportions in a recipe or compute gas mileage, markups and markdowns, or sales tax.

MATH BY ROTE

Maybe A's in math came easily to you in high school because you had a good memory and learned by rote. But most of the time you had no idea what you were doing. Passing courses by memory alone makes you feel like a fraud and as though you are skating on thin ice. You need to get your feet on the ground, become rooted, and understand what you are doing. Doing math on automatic and not *owning* your knowledge increases your anxiety.

Do you relate to all this fear and trembling when math is involved? What causes these people to be math-anxious? What causes *you* to be math-anxious? As a child, were you pressured to memorize the multiplication tables, tortured by flash cards? Were you left back a year? If you were a girl, were you discouraged about learning math? Did people in your life make you feel unimportant, stupid, or inferior to begin with? Can *you* recall the day of the traumatic scene? Do you remember a major event in which the Math Monster wrapped his hands around your throat? See if you can remember. It's time now to explore the origins and the intensity of your math anxiety.

3.. HOW MATH-ANXIOUS ARE YOU?

You'll probably find the roots of your math anxiety in your childhood. Like most small children, you started school with a sense of anticipation and a love of numbers ("Look, Ma, I can count to a hundred zillion!"), but along the way your love turned to loathing because math became so abstract, boring, and confusing, and you were labeled as The Boy (or Girl) Who Can't Do Math. You may have adopted some of your parents' and teachers' views and attitudes about math—and about your ability to do math. Our culture divides knowledge into territories, and children get labeled at a young age. Bobby is the "scientist," Louise is the "writer," and Anne is the "handy" one. Louis is the "fast" one, Jim is the "slow" one. (Well and good if you're Louis, but not so good if you're Jim.) Maybe you were strong in English, and so "they" assumed you were weak in math. Whatever happened to the Renaissance man (or woman)?

Are any of the following typical scenarios familiar to you?

In the fourth grade you were perfectly comfortable with math, but once you missed an answer and your teacher sent you to stand in the corner while your classmates snickered with pleasure. Or in the fifth grade, measles, scarlet fever, or strep throat kept you out of school and during that time your teacher introduced your class to long division, multiplication, or fractions. Even if you were absent only three weeks, you may have gotten lost in math for life. You came back to find that the worksheets, books, and blackboard were covered with squiggles that made no sense. You fell behind. Maybe in junior high you asked one or two questions but didn't quite get the answers you needed to fully understand. You fell behind. You decided math was not for you. Or maybe your father decided that numbers, along with hammers and power drills, were his domain, and sat you down every night to go over the obligatory math homework, causing a lot of screaming and tears. Perhaps, as in Vanessa's case, he called you "stupid." Or maybe a teacher became impatient with your questions and told you to be quiet and just listen.

Can you relate to any of the above situations?

I tell my students that an inexpensive, useful way of solving any problem applies to math anxiety: *observe it*. Pay close attention—*lots* of attention! You have to observe when math anxiety shows up, and ask yourself questions: Is it *people* who make you math-anxious? Is it *tasks* that must be performed that make you math-anxious? What memories get jogged when you're in a situation that produces math anxiety? What is the crucial element that triggers your anxiety? Is it the feeling of being tested, of being on the spot, or of reliving failure?

Here's what some of my students have written about math anxiety:

I remember that my problems started in the fourth grade. In grade school, I used to have arguments with my father. When he tried to help me, he always used "shortcuts." He just didn't see why I had to learn the long way.

I took bookkeeping at Katharine Gibbs, and I did fairly well because I could see a reason for it. Fortunately, in my first job no math was involved. In my second job I had to handle the department budget, and that was definitely a problem. I was sure I was never right.

You know now that you're not alone in your math anxiety, and you know just how many causes of math anxiety there are. The following quiz will help you to find the crucial features of your problem with math. You'll see what blind spots you have developed, and what area of your life is most affected.

HOW DOES MATH ANXIETY AFFECT YOUR LIFE?

YES   NO

CHILDHOOD

1. Do you panic when somebody asks you a question straight to your face? ____ ____

2. Do you think that it is essential to know your multiplication tables at the snap of your fingers? ____ ____

3. Do you have trouble with your sense of direction? ____ ____

4. Do you think there is one right way to do math? ____ ____

5. Do you have trouble standing up for yourself? ____ ____

6. Do you avoid conflict at all costs? ____ ____

7. Did teachers humiliate you? ____ ____

8. Do you compare yourself to friends and colleagues and always end up the less bright and educated one? ____ ____

SELF-IMAGE

1. Are you better at figuring things when it does not matter? ____ ____

2. Do you brood over past failures, real and imagined? ____ ____

YES   NO

3. Are you reading this book in secret? ____ ____

4. Do you believe that perfection will be around the corner once you are a math whiz? ____ ____

5. Do you say, "Yes, but," after somebody answers your questions? ____ ____

6. Do you have trouble describing yourself? ____ ____

7. Are you overly responsible and punctual? ____ ____

8. Do you hate lovers who love math? ____ ____

9. Are you the only person who knows about your math weakness? ____ ____

10. Do you think that happiness will be yours once you master math? ____ ____

LEARNING EXPERIENCE

1. Do you have absolutely no memories of math classes when you were a child? ____ ____

2. Are you uncomfortably certain that your brain is missing a crucial piece: the math lobe? ____ ____

3. Do you break into a cold sweat during a math test or a task involving math? ____ ____

4. Does the thought of a mathematical task make you want to leave town? ____ ____

5. Do you have memories of a blackboard full of squiggly symbols dancing before your eyes? ____ ____

YES   NO

6. Do you have a sense of doom around numbers that makes your body go numb? ____ ____

HEARTH AND HOME

1. Do you have trouble listening to a five-minute cocktail party conversation about numbers and statistics? ____ ____

2. Do your eyes glaze over when you read the results of your physical (blood pressure, cholesterol, triglycerides, etc.)? ____ ____

3. Do you panic over tax forms or instructions on how to use a machine? ____ ____

4. Do you flip past graphs and charts in newspapers and magazines? ____ ____

5. When on vacation, do you have difficulty handling foreign currency? ____ ____

6. Do you avoid tipping the hairdresser, cabdriver, and maid at a hotel for fear of being wrong (not because you are cheap)? ____ ____

7. Do you worry about America going metric? ____ ____

FAMILY TIES

1. Does math anxiety run in your family? ____ ____

2. Do you have an older sibling or a spouse who is a math genius and constantly reminds you of it? ____ ____

	YES	NO

3. If you were mother's (father's) little helper, did your own problems seem unimportant? ____ ____

4. In your family, do numbers come up only around tense money issues? ____ ____

5. Are you mortified when you make a mistake? ____ ____

6. Did people call you "stupid" when you were a child? ____ ____

7. Does your family keep secrets? ____ ____

8. Do you consider not knowing numbers a moral failing? ____ ____

9. Did one or both of your parents tell you that you would never amount to much? ____ ____

10. Does your brain blank out when somebody wants you to compute in front of him or her? ____ ____

11. Do you balance your checkbook with tears and gnashing of teeth? ____ ____

THE GREAT AMERICAN PASTIME: SHOPPING

1. Do you get nauseated when somebody explains pricing or discount or comparison shopping? ____ ____

2. Do you carry small bills (singles and fives) exclusively, to make sure you won't get cheated on change? ____ ____

3. Do you avoid challenging a sales clerk, even if you know the change is wrong? ____ ____

4. Do you take discounts on faith? ____ ____

                                              YES   NO

5. Do you tip without regard for the amount of the bill? ____ ____

6. Are there any buttons on your calculator you don't understand? ____ ____

THE DAILY GRIND: YOUR JOB

1. Have you passed up opportunities in order to avoid working with numbers and machines (computers)? ____ ____

2. Do you think knowing math will water down your creative talents and other skills? ____ ____

3. Do you feel stupid doing math in front of your boss or colleagues? ____ ____

4. If you solve a problem correctly, do you still feel like a fraud, asking yourself, "What about next time?" ____ ____

5. Do you leave memos unread if they contain statistics or a graph (or just a bunch of annoying numbers)? ____ ____

6. Do you worry about the horrible moment when you will be found out? ____ ____

7. Do you pretend to know math you don't understand? ____ ____

8. Do you leave the room when people discuss union business, pay raises, fringe benefits? ____ ____

9. Do you put off billing and other jobs involving numbers? ____ ____

SCORING

Give yourself 2 points for every "Yes," 0 points for every "No" or if the question does not apply to you, and 1 point if you feel undecided. Then add up your points. Check your total against the following:

80 points and up	Serious math anxiety
50 to 80 points	Medium math anxiety
20 to 50 points	Mild math anxiety

SERIOUS MATH ANXIETY

You are suffering a lot around numbers. Past experiences weigh heavily on you, and the faster you undo some of the myths and blocks, the better. You might profit from a patient teacher or tutor, or just a good friend to sit with you and talk through some of your reactions to math. Pay special attention to the Confidence-Building Techniques in Part III, returning to them whenever you need them.

MEDIUM MATH ANXIETY

In the past, you have been dealt a few blows to your self-esteem in selected areas of math. You can start applying the Confidence-Building Techniques in Part III and learning the basic math skills you may have missed in school, and you'll see that playing with numbers will make you feel a lot better. You'll profit from cultivating your favorite Confidence-Building Techniques and using them every time you falter. Just focus on staying on track, doing a little math every day, and applying your new skills immediately at the store, on the job, and at home.

MILD MATH ANXIETY

Your math anxiety is manageable at this point. Cheer yourself on, practice a lot, and let yourself enjoy some of the math facts and patterns that come easily to you. Your relationship to math is free of major obstacles. You probably just need a brush-up on some of your skills and greater ease in juggling numbers in your head. Maybe you're a parent or teacher with no math anxiety, but are reading this book for your students or your children. If this is the case, I hope this approach to math yields a few insights for you.

One of the first steps toward losing math anxiety is to treat your quiz score lightly. Your score is a loose, approximate indicator of your math anxiety. In our culture we already have too many numbers pinned on us that classify us and are used to make us feel unworthy. Yet the quiz does hold important information for you. Besides the total score, look at the clusters of ''Yes'' responses in the various sections. Do they occur in the job section or in the home section? Many people have different stances at home and at work. If you know about your areas of strength and weakness, you can start shifting positive attitudes and skills from one area to the other. For example:

LISA'S TEST

Lisa found out that she had almost no math anxiety at work, but her scores in the ''Childhood'' and ''Hearth and Home'' sections were high. Lisa and I discussed her difficulties, and she decided to change the situations at home that made her so insecure. One of them was the ''budget meetings'' with Tim, her number-whiz husband. (''Usually, when Tim tries to explain something to me that is math or has to do with numbers in any way, a veil comes over my eyes and brain and I get lost.'') Lisa and Tim rescheduled their meetings from the traditional late-night slot to weekend mornings. To support herself, Lisa decided to wear her work clothes and pull out her calculator; in short, she translated her business demeanor and math competence into her home. The actual and psychological effects of her strategy reintroduced a sense of balance into the situation, and Lisa could calmly and professionally discuss the budget with her husband.

JIM'S TEST

Jim found that his scores were all low except in "Learning Experience" and "The Daily Grind: Your Job." Every informal situation was easy for him because he could take his time and figure the math any way he wanted to. Only when he felt closely scrutinized did his hands get damp and his head start throbbing. After some strategy sessions with me, Jim learned to ask for the time and privacy at work that he needed to do his figuring. Billing turned from a nightmare into a routine chore. Jim also realized that as an adult he could handle learning and training in a more independent and assertive manner than he did as a child. He decided to go back to school, but on his own terms and at his own pace. He enrolled for one adult-education class and actually enjoyed himself.

MAPPING OUT
YOUR MATH ANXIETY

You want to find out about the nature of your math anxiety: Is it a person, a place, or specific math tasks that get to you? Is it the feeling of being tested, the feeling of being on the spot, of reliving failure? Sheila Tobias, the author of *Overcoming Math Anxiety,* suggests a wonderful technique to help you focus on your anxiety: a "math autobiography" that helps you pay attention to your own story, its details and protagonists. Don't skip this exercise. You'll find it especially useful after taking the quiz.

Writing your math autobiography focuses and revives your memories, clarifying the history of your math anxiety and the events and other people in your life that helped create the problem. Before you start writing your own, it will help to read excerpts from two autobiographies to see how other math-anxious people have explored their pasts. This is useful because our own memories often hide the most painful moments or feelings from us. Reading a similar event in somebody else's story helps eliminate your blind spots. Here's what Jenny and Belinda wrote in their autobiographies.

JENNY'S AUTOBIOGRAPHY

First and second grade were not a problem. In second grade we started a more difficult subtraction, and I think I was confused for a while, but eventually I understood it. In third grade we started multiplication. The material was not difficult and included lots of memorization. We would memorize our times tables, and then the teacher would flash cards and we had to say the answers very fast. Fourth grade and fifth were difficult in math because of the introduction of fractions, of which I could make no sense.

My parents were also having problems, and by the time I was in the fifth grade, they were separated. Eventually they divorced. Sometimes I think that might have had some influence on my problem, but I hate to make excuses. The rest of my junior high and high school years were difficult when it came to math. Anything having to do with percentages, fractions, and eventually algebra gave me the shudders. For a long time I would try to avoid taking math classes unless they were required, just because I hated the feeling of being lost in class. Now I'm in college and still struggling with math.

Chances are I'll have to seek help from a tutor, which is fine with me. I know I need individual attention when it comes to math. Maybe all I have to do is fill in the missing link—fractions and percentages—and the rest will come easier.

BELINDA'S AUTOBIOGRAPHY

Math, to me, means years of suffering. This problem has caused me to feel inferior in many ways, most of all because I didn't feel competent to go on to get a master's degree.

In my present job, I write job descriptions for people two and three levels higher on the corporate ladder than I am. I also take active and constructive roles in the executive appraisal process for these people. Yet I have a block when it comes to math.

My first memory of this nightmare is of when I was nine years old and getting ready for a math quiz. The passing grade on this quiz was 85. I spent the worst night, tossing, turning, and dreaming about failing that quiz, and fail I did. This was the second quiz of the semester, and I barely made 85 on the first one. The teacher's sermon about "dummies" kept replaying itself in my head. When I closed my eyes I could see her face, looking cheerful when a student failed and barely saying anything when she saw a passing grade. I was going to fail, I just knew it, and my mother was no help at all. When I told her the first week that I wanted to change math classes, she asked why. After I explained that the teacher always went over the problems too fast and never gave the class a chance to copy the examples from the blackboard, she said, "Oh, you just write too slowly."

One day I wanted to stay after class and ask a question. The teacher glared at me, and I slid lower in my seat before finally asking her if she could go over one problem with me. She said, "Belinda, you had better get on the ball and hear what I am saying, copy the examples from the board, and memorize them." I said, "I'm sorry, but I just don't understand how it all works." I was told, "You'd better start to understand, or I will fail you and you'll be right back here with me next year. Try to be more like your brother. He never fails a quiz and can keep up with me in the math text."

Quiz day! I was shaking, and my hands were clammy. I had a pocket full of tissues. I sat in the back row for this one, and did not talk unless I was spoken to by name. I missed one problem too many and scored 65, not 85. My name was called along with four others. We marched to the corner. Dunce! I found a corner and stood with my back to the class. The rest of the class laughed and jeered at us, the teacher chiming in as well. With twenty minutes of class time left, the remainder of the class went over all of the questions and answers. The "Dunce" group, who needed this drill the most, was denied it.

I will never forget that teacher's name, and I haven't spoken to her since the last day of that math class. I wanted to tell her to go to hell. That experience never left me. It was relived on a regular basis that entire summer. My mother was ready to enroll me in a summer math workshop, but my dad told her to "give it a rest."

There are many memorable math occurrences in my life, but this is the one I remember best.

YOUR MATH AUTOBIOGRAPHY

Ready? Now take a blank sheet of paper, start writing, and after you've finished, put your autobiography aside for a day or two. The time lag will give you more distance and a chance at objectivity.

When you look over your autobiography in a few days, start to analyze it. How can you analyze your own autobiography? It's a good idea to read it with a few guiding questions in your mind. You might concentrate on the following:

Who are the people in your story that are mentioned most frequently? Parents? Siblings? Teachers? Friends? Spouses? Colleagues at work? Nobody?

Are there *crucial events* that made math scary for you? Were you humiliated? Were you left back in school?

Was any decision about your *education or career* determined (or influenced) by your math anxiety?

Has math anxiety affected your *self-image* and lowered your self-confidence in other areas of your life?

Is all the *action* in your math autobiography *far in the past,* or is it a *continuing saga?*

After you have pondered and answered these questions, you will certainly have a good picture of your math anxiety and will be well on your way to overcoming it. You're ready to start keeping a math journal.

KEEP A MATH JOURNAL

Since writing is one of the most effective ways to make sense of things, keep a small "math journal" to carry with you at all times. You can heighten your awareness by noting the circumstances in which your anxiety is triggered. Who triggers it? Where is it triggered? In restaurants? At work? In front of other people? Maybe you never have trouble alone, but restaurant situations with other people intimidate you. Maybe figuring the costs of a trip or a weekend share, or balancing your budget unleashes the wrath of the Math Monster. When you actually write out what these external/old/hostile/obsolete voices inside your head are telling you, they tend to vanish like vampires in sunlight. It may also help to record incidents that cause you math anxiety, along with physical symptoms of discomfort, as well as any coping strategies you find useful. Always note who was there, where you were, how you felt. Analyzing this record later will reveal a wealth of patterns.

TELL SOMEONE
ABOUT YOUR MATH ANXIETY

To lift the burden of secrecy, reveal your math anxiety to some-one important in your life. Choose a person who will be totally supportive of you and who will understand that math anxiety takes a terrible toll on your life. Keeping secrets that involve shame or embarrassment creates a burden and interferes with the functioning of your mind. Shameful secrets further thinking that doesn't flow smoothly, that peters out, that doesn't arrive at an answer; you cannot think straight or be open; you cannot listen or respond emotionally. If something touches on your secrecy is-sue, even tangentially, your defense mechanisms shoot up, and thoughts are derailed; you go blank or become confused or angry or flustered. To tell even one person your story relieves some of that stress. (This is one of the rationales behind confession!)

When my student Laura told her boss about her math anxiety, he responded—to her surprise—sympathetically. ''Why have you suffered in silence so long?'' He bought her a gimmicky calcu-lator and helped her with some of her homework from my course. Later he sent her to an accountants' training program. Laura improved dramatically at her job, and not only in the math area. She felt free.

Thus far we've examined the nature and origins of math anxiety. You now know that you have plenty of company, and that it's a national problem. You've taken the quiz and have some under-standing of the situations in which you feel most math-anxious. You know when your math anxiety began. What follows in Part II is a perspective on the cultural conspiracy that created and constantly reinforces your math anxiety. You'll read about our culture's math myths, and the blocks you've erected in your re-sponse to these myths.

II...
THE CULTURAL
CONSPIRACY

4 .. MATH MYTHS

Although your math anxiety probably began with a person or an event in your childhood, a cultural conspiracy reinforces that anxiety. Part of that conspiracy is your belief in frequently held myths that inhibit your natural performance and common sense. These myths, powerful because they are constantly echoed by parents, teachers, math books, the media, and, of course, your own conscience, cut you off from your innate intelligence and limit your natural curiosity.

Do any or several of these myths that limit your approach to math ghost through your mind?

1. *Math has to be approached with cold logic.*
2. *There is a formula to be remembered for each problem.*
3. *Math performance mirrors intelligence.*
4. *Math is for men.*
5. *To be poor in math is a moral failing.*
6. *If you are creative, math will spoil your talent.*
7. *You must know how you got the answer.*
8. *There is one best way to do each math problem.*
9. *It's always important to get the answer exactly right.*
10. *It's bad to count on your fingers.*
11. *All mathematicians do problems quickly in their heads.*
12. *Math requires a good memory.*
13. *Math is done by working intensely until the problem is solved.*
14. *Women can't think straight.*

If you identify with six or more of these myths, you are allowing yourself to be oppressed by phantoms.

Let's refute each one of them.

1. MATH HAS TO BE APPROACHED WITH COLD LOGIC

Math is often thought of as a self-contained set of procedures and rules, which you must approach in a coldly logical way. Not so. Unless you understand math's underpinnings and connections, you won't know how to use the rules. Actually, most of the math you need in daily life requires few sophisticated skills or cold logic; what you need instead is plenty of common sense and intuition. When you *limit* yourself to a strictly logical approach to math, you miss out on your own perceptions and the "first thoughts" that Natalie Goldberg describes in *Writing Down the Bones*: "First thoughts have tremendous energy. It is the way the mind flashes on something." I believe this describes the sort of intuition that is important in math. (See "Brick Arithmetic," page 133.)

2. THERE IS A FORMULA TO BE REMEMBERED FOR EACH PROBLEM

The math we encounter every day rarely requires one special formula to solve the problem. If you are competent in math, you know that you will arrive at the answer sooner or later. You don't give up when you can't spot the answer at once. You give yourself one chance after another to find the answer. You restate the problem in your own words, imagine yourself in the situation, and ask yourself:

"How can I explain it?"

"Can I draw it?"

"Can I do it another way?"

"What else can I do?"

"Can I omit part of it?"

"Can I do the first part?"

"Have I seen anything like this before?"

Seven different methods may prove wrong, so you try for the eighth time. Right!

It is this "thinking on your feet" that makes you successful in math—not an ability to reach into a storehouse of dead facts and pluck out a number or formula. You should only memorize a math formula when you understand the reason for it. Half-digested rules like "turn a fraction upside down and multiply" only do damage if they are unconnected to reality.

3. MATH PERFORMANCE MIRRORS INTELLIGENCE

When people connect math performance to intelligence, they equate intelligence with an ability to reason abstractly. They are wrong on both counts. Math competence, and particularly mathematical inventiveness, requires good doses of intuition and imagination. Intelligence itself is a more complex phenomenon than just reasoning ability. Unfortunately, our culture leans toward left-brain, sequential, analytical thinking, often undervaluing artistic, practical, and commonsense-type intelligence.

We tend to teach math according to these preferences, in a sterile and routine-oriented manner. But problems in math, surprisingly to most people, can be approached in different ways. Scholars have identified hundreds of different learning styles, but our schools do not inform us about how we think, how we solve problems, or how we process information. We tend to take our minds for granted. *Most people spend more time learning about their car or their Cuisinart than they do learning about their own minds.* Styles of learning have to do with sensory perception and temperament. Do you look more than you listen? Are you an active or passive learner? In addition, the theories about left-brain/right-brain thinking have shed new light on people's processing of ideas.

You will find out more about thinking and learning styles (and how to optimize your own) in the books listed under "The brain, learning, and consciousness" in the Recommended Reading suggestions on page 221.

I think a visual learning style is a great advantage in learning math. It helps to imagine or visualize the situation and draw a picture of it. It also helps to see numbers clearly in your mind because you can manipulate them on your internal movie screen. I tend to picture a large white movie screen on which I put my numbers, printing the ones on which I want to focus in big pink neon digits, and they stay there cheerfully. We're going to refer to this when we do mental arithmetic, which is one very useful skill. It is important because people actually make fewer mistakes figuring things in their heads than writing them down, and it allows you to think on your feet in meetings and negotiations.

The "intelligence myth" is peculiar to Western culture. Americans think of mathematics as an esoteric skill, like perfect pitch, that some people have and others do not, but when social scientists investigated the reasons Japanese and Chinese children were outperforming American children in math, they discovered that the disparity in math achievement had nothing to do with superior teaching techniques or a "talent" for math. Instead, Chinese and Japanese adults believe that math ability is pretty evenly distributed among all children, and that hard work is what makes the difference between math competence and failure.

4. MATH IS FOR MEN

Math, more than most other fields, suffers from a gender myth. Studies "prove" over and over that boys are better in math than girls. An obvious bias prompts these studies. (I wonder who pays for all of them?)

Most of us are hypnotized by these studies. Trying to show gender-related differences in a field that is steeped in gender-related prejudices is a rather easy task. At this moment in history (women have entered the field only for one or two generations), measuring the average performance of all girls versus all boys is less important than encouraging girls who are gifted and/or interested in math.

Developmental theory has established men's experience and competence as a baseline against which both men's and women's development is then judged, often to the detriment or misreading of women. Let me quote from *Women's Ways of Knowing: The Development of Self, Voice and Mind* by Mary Field Belenky, Blythe McVicker Clinchy, Nancy Rule Goldberger, and Jill Mattuck Tarule:

> Nowhere is the pattern of using male experience to define the human experience seen more clearly than in models of intellectual development. The mental processes that are involved in considering the abstract and the impersonal have been labeled "thinking" and are attributed primarily to men, while those that deal with the personal and interpersonal fall under the rubric of "emotions" and are largely relegated to women. As dichotomous "either/or thinking" is so common in our culture and as we tend to view human beings as closed systems, the expenditure of energy in one part of the system has been seen inevitably to lead to depletion elsewhere.

Math has always been a stumbling block for women. They have had trouble overcoming age-old pressure, pressure that tries to keep them at home and unknowing. Women traditionally took little part in the development of math. The most famous female mathematician of antiquity, Hypatia of Alexandria, was brutally murdered by an envious male. Women in Europe in the Middle Ages who were knowledgeable and brave enough to show their knowledge often faced prosecution as witches, and execution.

From the days of Greek civilization, math has always been invented, practiced, and studied in the relative isolation of secret societies or universities, and always in exclusively male elite groups.

Many women feel that they're intruding on male territory or breaking a taboo when they do math. And for good reason! Although schools today teach girls as much as they teach boys, and most universities have opened their doors to women quite freely, there is still hesitation on the part of these universities to accept women in the top ranks of the mathematics department, as well as hesitation on the part of women to go into that lion's den. Women lack visible role models in math and science. (Yes, there was Madame Curie, but she is often the only example ever cited.)

Happily, the myth that math is for men is being challenged and eroded every day. Women are moving in and up in math, science, and computer-related professions and careers in growing numbers, as are blacks, Hispanics, and other groups. Math has ironically become a golden road to opportunity. In the recent movie *Stand and Deliver*, Jaime Escalante, a high school math teacher, teaches poor Hispanic kids calculus because he sees math as the "great equalizer." The movie, based on a true story, shows that once women and minorities claim their rightful place in the math and science programs of our universities, they will have a good chance of riding out the discrimination and harassment by solid achievement.

5. TO BE POOR IN MATH
IS A MORAL FAILING

When people cannot do a math problem, their anxiety level often rises to life-or-death proportions. As a friend of mine said, "It's like Judgment Day." Math incompetence is equated with worthlessness, an attitude that may have evolved from the Puritan work ethic, in which sloth is one of the Seven Deadly Sins. Math is seen as the only subject in which you're either right or wrong— "right," in our culture, meaning morally correct. The language takes on a very fundamentalist tone, but shame or moralizing have no place in assessing knowledge.

6. IF YOU ARE CREATIVE,
MATH WILL SPOIL YOUR TALENT

"Creative" people—writers, artists, dancers, and musicians— often believe their brains function in a special way that allows them to be creative. And they worry that math, which they see as pedantic, gray, and dull, will contaminate, hinder, or even kill their creativity. This attitude is similar to the attitude of some creative people who refuse psychotherapy on the basis of their belief that their depression or neurosis enhances their creativity, and that if they conquer it, their talent, diminished, will fly out the window, leaving in its wake a dull-witted bore.

Yet history shows that many individuals have harbored various talents without any conflict. Leonardo da Vinci was both a gifted artist and a great mathematician/technician. I have a fiction-writer friend who reads *Scientific American* and mathematical puzzle books when she has writer's block. She says that once her brain has fed on all those neat little numbers, it is ready for the next chapter of her novel. Our minds are vast and flexible, and we do ourselves an injustice when we subscribe to a "scarcity model" for our mental activity. The energy and mind-power you need to do math is not subtracted from your creative talent. In fact, new ways of seeing, thinking, and reasoning may well infuse your talent with new ideas.

7. YOU MUST KNOW HOW YOU GOT THE ANSWER

People who believe this myth have bad memories from childhood. They remember having a right-brain, intuitive way of doing math, where pictures and numbers dance around and seem to fall into place. But even though the *correct* solution emerges from their vision, a teacher will ask, "How did you get that answer?" The teacher, insisting on a formula, refuses to look at a discussion of pictures and visions. If you have tried to capture your process and you look up at the ceiling, the teacher is likely to warn, "The answer is not on the ceiling." Your classmates giggle and sneer. Actually, the child who looks up at the ceiling is a visually oriented child who is trying to recall the *image* in her mind. If you were such a child, later in life you may persistently doubt your answers because they arrive in an unorthodox way.

8. THERE IS ONE BEST WAY TO DO EACH MATH PROBLEM

There are many ways to do any math problem, and *your learning style, the way you process information, is the way you'll do it best.* Is it easiest for you to learn by reading and looking? Is it easiest to listen? Is it easiest to "go through the motions," actually to use your body to learn? You may be quite unaware of which of the three you prefer. All of us, of course, use all of them, but we tend to have a dominant learning style. Have you noticed that some people take notes while others repeat things out loud?

When right-brain methods are emphasized in learning math, you imagine the setting in which the math problem occurs. You eliminate anxiety, narrowness, and rote memory (functions of the left brain) by using common sense, intuition, and humor. If you *put yourself into* the math problem, by knowing who the protagonists are, what kind of shoes they wear, and the color of the cash register, you can make use of the *relational* mode. Traditional math and science often look down on personalized process and knowledge.

9. IT'S ALWAYS IMPORTANT
TO GET THE ANSWER EXACTLY RIGHT

The people who are often most admired as math whizzes are those who bounce numbers around generously, rounding, estimating, getting the bigger picture in seconds while their colleagues are still recovering from the influx of data and reaching for their trusty calculators.

Budget meetings and negotiations are cases in point. The exact share of your department's salary expenses is not half as important as the *rough* percentage and *roughly* how much more you want.

10. IT'S BAD TO
COUNT ON YOUR FINGERS

It's not "bad" to count on your fingers. Little children, for instance, learn a lot more when they are allowed to count on their fingers or count objects rather than moving too quickly to abstract numbers. The number symbol 5 means nothing to a small child, but five M&M's on a plate mean a lot. If you have a geometry problem and you fold paper, draw, and use your hands, your thinking patterns and nervous system respond. I am convinced that people who are adept with an abacus know the math with their hands as well as their heads. The same is true for Chisanbop, a finger-math technique that has yielded incredible results in American classrooms.

11. ALL MATHEMATICIANS RELY EXCLUSIVELY ON ABSTRACT THINKING

Mathematicians do not rely on abstract thinking at all times. They do not always sit at their desks with sharpened pencils. They're not shy about getting all the help they need. They discuss their work with colleagues. They turn to every tool that can assist them. They use computer graphics to give them visual representations of what they're thinking about. Mathematicians I know who are involved in high-level math have no hesitation about pasting and gluing models—drawing, imagining, personalizing—to make it concrete and real.

12. MATH REQUIRES A GOOD MEMORY

A good memory always comes in handy, but is less crucial in math than in other subjects. What math requires is a high level of concentration. Math is, in fact, 90 percent concentration and 10 percent ability. Any strategy that makes you focus—meditation, memory techniques, crossword puzzles, language skills—will pay off in math.

13. MATH IS DONE BY WORKING
INTENSELY UNTIL THE PROBLEM IS SOLVED

Not necessarily so. When you study math (oh, yes, it's coming up in Part IV), you should take five-to-ten-minute breaks (at least) every thirty minutes or so. How often depends on your own natural rhythm, as well as the difficulty of the subject and your interest in it. If you organize yourself this way, you'll give your mind a chance to rest and to sort out the information you've been processing. Often, when you let a pesky problem alone, the solution will show up unexpectedly an hour or a day later.

14. WOMEN CAN'T THINK STRAIGHT

One of the reasons I decided to write this book was to make women aware. Many women walk around thinking they are not very bright or at least that they lack the ability for good, thorough, analytical thinking.

Little do they remember or connect this so-called lack with incidents of early abuse or the constant chiseling away of self-confidence by subtle discrimination, derogatory remarks, and the constant sexist drivel that the media perpetrate and most of us don't notice anymore.

Many women, on learning that abuse and intimidation leave traces on the mind and not just on the soul, start feeling relieved and hopeful. If they can get the old, unfinished business out of the way, maybe they can actually think. There is actually nothing wrong with them. The effect of abuse and intimidation on thought processes is demonstrable; I call it "sideways thinking."

Sideways thinking characterizes many victims, many of them women and members of minorities. It makes the thinker believe she is incapable of reaching logical conclusions, while all the while she has been *taught* not to let a thought complete itself. A victim's way of thinking is circular. *Yet people can climb out of the victim's stance.* They can take charge. Thinking changes. Power is thinking that leads to conclusions. I was hurt, yes, but I am not dumb, I am brilliant.

I do know about the terrible effects of shock and abuse on thinking. In my work with the Rape Intervention Program at St. Luke's–Roosevelt Hospital Center in New York, I meet women who function well enough on the surface; they smile at the doctor, say "thank you" to the nurse. Yet their thinking is short-circuited, obsessively circling the crucial moment: "Maybe if I had had my key in my hand"; "Maybe if I hadn't gone to the movies"; "Maybe if I had insisted that a friend go home with me"; and on and on, in smaller and smaller circles.

I get up in the middle of the night to help women break that circle in the first few hours after the crime has been committed, before that deadly spiral thinking pattern has had time to root and spread to other areas of that woman's life.

Many people I meet in my classes and in private lessons have settled into sideways thinking, and have to work hard to undo it. They have to dig out the trauma, the hurt, the terrible feelings of being oppressed and unjustly treated. First they must face their pain and anger, then go out and use that emotion instead of letting it fester. There is great energy in anger, and once it is in the open, good things can emerge.

Second, they must work to change their self-image, realizing that all the doubts, the double takes, the depressions, the feelings of stupidity are not intrinsically theirs, but were induced from the outside. If doubt and depression can be induced, they can be erased; then the former victims can begin to realize that their minds are good minds and that they can change from feeling dumb, limited, and unanalytical to feeling smart, sharp, and on top of things.

I have witnessed such transformations, and it is a great day when clarity strikes and issues can be clarified, separated, and dealt with.

These myths play havoc with our minds, the more so the less we are aware of them. When you read newspapers, listen to people talk, or watch a talk show about education on TV, notice how many of these myths turn up in the protagonists' minds. Catching the myths every time you are ready to fall for them helps to immunize you against their immediate effects.

You may, however, have made yourself immune in a less productive way by erecting defenses that nicely match the message of each myth that affects you. If you are female, you may have bought into the "women can't think straight" myth by reasoning that "I am a woman; women can't think straight; ergo, I can't think straight." Perfectly logical, but nonetheless incorrect. If you repeat this many times to yourself and have it reinforced by the people around you, those initially innocuous little short-circuit reasonings have grown into big, solid defenses or math blocks. Many people I meet refer to them that way. A friend recently said, "I have a math block as big as a house." In the next chapter we take a close look at some of the common math blocks and how they dictate behavior in "math situations."

5.. MATH BLOCKS

A block is an obstacle to the free-flowing, capable functioning of your mind that stands between you and math competence. Math blocks can be so solid that you tune out the voice that explains math to you, and your vision blurs at the sight of numbers on the page.

Blocks cannot be blasted away overnight, as if with dynamite. We have built them over the years for a reason, but in this book we will learn to use our minds in new ways and remind ourselves how strong, efficient, and smart we are in other areas of our lives. In this way we will either chisel away at the blocks or walk around them.

Math blocks, defenses to fend off math anxiety, often lead to avoidance of real thinking and learning. Like writer's block or other performance blocks, math blocks have plagued many talented and capable people I have met. Blocks are normally expressed in the idea "I can't. No way." Nervous questions of self-esteem hang there ominously. Will I measure up? Will people laugh at me? Will I freeze in mid-performance? Will I fail again? "I have a math block," Ann said. Then she sat back, relieved at having revealed her problem and sure that possessing a block was all the explanation we needed.

"Did you bring it with you?" I said, smiling, making her imagine her "block." Ann burst out laughing.

Blocks fall into patterns, typical resistances to numbers and math that have their beginnings in how we grew up and how we look at life. I have found seven prevalent math blocks. Find the ones you recognize and identify with from the list that follows.

1. PERFECTIONISM
("Yes, but . . .")

You diminish your accomplishments with pretensions toward lofty goals. Your attitude stands in the way of acknowledging progress. You are so focused on the "right method" and the "right answer" that the process required to get to the right answer makes you uncomfortable. It's hard for you to appreciate your competence or progress.

TYPICAL BEHAVIOR: At the end of a math problem, you rush on to the next one, saying, "Yes, that was easy, but what if the numbers are harder?" "I'm correct, but I didn't do it mathematically." "I can do algebra, but my father got an A in calculus."

2. REGRESSION
("Little Me")

Remaining a child makes you overvalue authority, makes you gullible, and lets you off the hook. Math is a human endeavor like any other, and was invented to be used on simple, real-life problems. If you play "Little Me," you underplay the powers of your mind, leaving a lot of the work and the rewards to others—those who know.

TYPICAL BEHAVIOR: During a math test, you are likely to give up too early. You let others do *your* work, avoid promotions that challenge you mathematically, and you choose careers that do not involve taking math courses or dealing with math on the job. These "negative career choices" are rarely satisfying.

3. INFERIORITY
("I'm stupid.")

You assume you are less bright than others. This attitude makes learning and doing math twice as hard as it needs to be, since math is often seen as a measure of intelligence. You may be just as frightened of discovering how bright you are as of the opposite. The "I'm stupid" attitude is learned early within the family. Being cast as the "stupid child" has a diminishing effect and actually can stunt development if the child does not get other positive feedback. These early messages are hard to give up.

TYPICAL BEHAVIOR: When doing math, you are likely to take a backseat, admire others, and almost expect not to learn.

4. PASSIVITY
("What's the formula?")

You have a strong sense that math is an immutable, mysterious, none-too-sensible collection of formulas, and you believe that if you had a good enough memory you could handle any math. You do not rely on your mind in a given circumstance. You have learned early to distrust your mind's output.

TYPICAL BEHAVIOR: After reading a problem, you are likely to ask, "Do I multiply or divide?" or feverishly try to remember that formula from seventh grade. In math as in other situations, you say or imply, "Help me."

5. TERRITORIAL
("I'm the creative type.")

You were typecast early as a "good English student," an artist, an athlete—anything but a "math-head"—by a society or family that divided the areas of work and achievement into territories. The power of family pronouncements keeps you from trespassing into taboo territory, implying that you might pollute the creative wellsprings of your talent and dry up your thinking.

TYPICAL BEHAVIOR: You are likely to avoid math completely, and are unwilling to find out how good your math can be.

6. CONTEMPT
("I'm not a nerd.")

You hold similar beliefs to the "creative" person, yet the emphasis is on distance *and* contempt. Dr. Paulos hits the nail on the head in *Innumeracy*, when he says, "Mathematics is thought to be cold, since it deals with abstraction and not with flesh and blood." You see math only as tedious manipulation of numbers and symbols, and mathematicians as bores with glasses and pale faces who do long division for fun on Saturday nights.

TYPICAL BEHAVIOR: You close your eyes to the interesting people around you who *can* do math and like it (they *do* exist). You avoid situations involving numbers, and certainly you would never take a math class or participate in a conversation involving numbers.

7. GENDER
("Nice girls don't do math.")

You buy into the myth that women have less talent for math and should not worry their pretty little heads over it (for their own good, of course).

TYPICAL BEHAVIOR: For you, the brainwashing began at an early age. Most parents, consciously or unconsciously, have different expectations of boys and girls. They tend to expect that their sons succeed and that their daughters try hard and smile a lot. All is not lost, however. The *New York Times* (not the most feminist paper around!) remarked on Saturday, July 1, 1989: "Gender Gap in Aptitude Tests Is Narrowing." Girls are catching up. They have been closing the gap slowly but surely. Our heredity has not changed since the sixties, but *attitudes* have. Younger women believe more strongly that they have the brainpower to do what they need to do. If you lived through the earlier, more inhibiting decades, take heart, look around, and, next time you have a math question, ask a woman.

You *can* knock down those math blocks and shake off those math myths. Keep going; in Part III you'll find some Confidence-Building Techniques that will help you to build your self-esteem before you start Easymath in Part IV. I'll teach you how to handle your panic. Learning these techniques and getting a grip on basic principles will not only give you new confidence, but will reverse your math anxiety entirely.

III...
DEALING WITH
YOUR ANXIETY

6 .. CONFIDENCE-BUILDING TECHNIQUES

Math anxiety does not disappear overnight, from one big "Aha!" experience. A big "Aha!" may have popped into your mind after you mapped out your math anxiety by taking the quiz, writing your autobiography, and relating to myths and blocks. True, understanding and acknowledging your math anxiety is a first step, but to reduce your anxiety and approach the math in Part IV with curiosity and courage, you need to build confidence in your daily life. You want to be ready when numbers confront you in your office, in a store, or during an employment exam. At these perilous times you'll have to walk yourself through the challenge, hold your own hand, and be your own best friend.

We tend to lose our cool when we have big goals. The more we push ourselves, the more others push us, the more ambitious we are, the more our craziness and anxiety build up. Thousands of years ago, the Chinese sage Chuang Tzu wrote about this in the following poem:

THE NEED TO WIN

When an archer is shooting for nothing
He has all his skill.
If he shoots for a brass buckle
He is already nervous.
If he shoots for a prize of gold

He goes blind
Or sees two targets—
He is out of his mind!

His skill has not changed. But the prize
Divides him. He cares.
He thinks more of winning
Than of shooting—
And the need to win
Drains him of power.

HOW THE CONFIDENCE-BUILDING TECHNIQUES WORK

How do you knock down the blocks that stand between you and your math? My clients use Confidence-Building Techniques that create a productive inner climate to keep them on track and to get them *back* on track when they fall off in moments of panic. These techniques, which take only a few minutes each, change your mood and mind so that anxiety dissolves and you can allow yourself to be cheerful, optimistic, and productive.

The Confidence-Building Techniques are a distillation of the most useful ideas and techniques that I've learned in twenty years of teaching math and helping friends and clients over the math-anxiety hurdle. First, become familiar with them. Read this chapter quickly (don't try to absorb the ideas all at once), then go back to the beginning and reread slowly. The techniques range from the most imaginative to the extremely down-to-earth. Even if you are skeptical, don't skip any, because each one helps in a different way. Suspend your disbelief and try them out. You should use these techniques every day as ''preventive mainte-nance'' to keep yourself calm and balanced. Then, when your anxiety is especially severe *(PANIC!)*, a combination of three techniques that form your Panic Kit will reduce the symptoms of your anxiety.

1. BREATHING FOR RELAXATION

Through proper breathing, you can put yourself in a state of physical relaxation. When you're anxious, you hyperventilate, so breathing correctly is important. The first step is to learn to relax your body through use of a systematic method, of which several are available. Some people favor yoga. Some people use meditation. For immediate use in panicky situations I recommend "Square Breathing."

SQUARE BREATHING

Square breathing uses four phases of breathing: exhalation, hold, inhalation, hold. It slows down your worry-mind and gives you more oxygen, and the counting keeps you busy while your system slows down. The rhythm is 4-4-4-4.

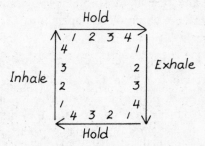

As an alternative to breathing exercises, if you're the type who prefers activity, running, aerobic dancing, yoga, and jazz dancing can all help to balance your system, increasing your energy and reducing worry.

2. AFFIRMATIONS

Old messages are strong, and what they tell us can make us or break us. Remember, you're still battling all those old negative messages. Learning new things triggers all the pleasant and unpleasant memories from school. Martha Friedman, writing in her book *Overcoming the Fear of Success* (Warner Books, 1980), about the negative messages from the past that continue to affect our lives, captures it perfectly:

> [When] we have internalized the messages of our parents and siblings, we move on to our second set of critics, our teachers, those whose job it is to further program us for success or failure by branding us with letters that run from A to F. . . . The manner in which we respond to negative criticism is a clue to the level of our self-esteem. If we harbor a feeling of inadequacy, negative criticism can wipe us out. Most of us carry with us too many internalized low-esteem messages from the past, negative things our parents or siblings or teachers or schoolday peers said to us. Even one negative remark may trigger any one of these messages, leading us to feel rampant self-doubt about our competency.

Your inner voice needs new songs and new sentences to maneuver you through tough spots and to help keep you afloat through the day. Affirmations are wonderful for dealing with all those old negative messages. On first reading, the following list of affirmations may seem a bit egotistic, yet they have nothing to do with bragging. When you practice affirmations, it's like playing powerful new music over scratchy old tapes. Read them over slowly; then reread them and pick an affirmation that jolts you pleasantly, that you deeply wish were true. Work with that one affirmation daily until it has become comfortable and natural.

I learn easily and steadily.
I give myself credit for everything I know.
I give my mind free reign to learn.
It's amazing what I can do.
My brain is working steadily, whether I study or not—whether
*   I'm asleep or awake.*
I am smart enough to learn anything I truly want to learn.
Being easy on myself really speeds me up.
Numbers are harmless little things.
I will go slowly, one step at a time.
I look at the pages of the math book and I see them clearly.
I read every word and every sentence and I take my time.
I look at symbols and I think about what they mean.
People have done math for thousands of years. I can do math,
*   too.*
Math is simply another way of saying things.
I am courageous.
I have my permission to try the impossible.
I am brilliant.
I am allowed to go far.

In my workshop, the affirmation "I am brilliant" jolts more
people than any other. Say it a few times. Say it out loud. You
may be concerned with honesty and feel that you are lying, but
affirmations are not concerned with the way you feel now; they
are statements of faith.

Go on, say it (in a whisper, if you like).

I am brilliant.

Many of you will respond by saying you're not Einstein, and,
in fact, the very last thing in the world that you are is brilliant.
Wait.
Say it again.

I am brilliant.

This is a great affirmation. See how it shines. What does *brilliant* really mean? Beyond the narrow intelligence that you're thinking of when you protest, it means bright, shiny, vivid, luminous, radiant, full of light.

Use it for a few days. Close your eyes while you say it and try to picture it. Write it across the top of your notebook pages, your daily agenda, the paper napkin in a restaurant. We need to keep our affirmations in the forefront of our minds so that we don't get lost in old habits and patterns. Return to the affirmations often. You may need different ones in the various phases of your mathematical adventure. Use one steadily for several days or weeks at a time, until it feels fully comfortable. Then choose another one, one that moves you in a new direction you want to go.

One nice way to use affirmations is to record the ones you have chosen on a cassette recorder. Say each one three times, pausing to allow them to sink in. Listen to this cassette before going to sleep. It might even inspire your dreams.

In my workshop, I asked Helen to say, "I am brilliant." As I expected, Helen protested, shaking her head. No, she could never describe herself in such a pompous way. But finally she agreed. She would say "I am brilliant" even if she blushed. "I am brilliant." "I am brilliant." "I am brilliant." Suddenly, after saying it several times, she sat up tall and smiled. "I like that. I like that feeling of being smart and shiny and light." The following week she admitted that by saying it many times a day, she had gotten through a difficult week at her job.

CARLA'S STORY

"My inner voice could use sweetening," Carla told me after class. "I still tell myself the same old ugly things that I always did."

"What do you want your voice to say?" I asked her, and she

fell silent. She had not thought about it. Unless you have a three-dimensional color picture of what you want to happen, your goal will be harder to achieve.

I asked Carla to imagine herself as delightfully competent in math. Once she was able to visualize her competence in a situation involving math, she straightened up and took a deep breath, and took on an expression somewhere between serenity and great determination.

"What is 80% of 2,000?" I asked her.

She paused, then calmly said, "Eighty percent of 2,000 is—wait—10% of 2,000 is 200—80% is 1,600." Carla had moved into a different mode in terms of thinking and feeling.

If you prefer another affirmation to "I am brilliant," then pick one and use it to further your goals. If you have trouble reading a page with numbers on it, you need the affirmation

Numbers are harmless little things.

or

I look at the pages of the math book
and I see them clearly.

Next, follow the directions for using an affirmation:

- Use it for a few days.
- Close your eyes while you say it, and try to picture it.
- Write it across the top of your notebook pages, your daily agenda, the paper napkin in a restaurant.
- Say it three times, pausing, to allow the affirmation to sink in.
- Speak it on tape to listen to before going to sleep.
- Keep your affirmation in the front of your mind and return to it often.

In a week or so, you might want to try another one, but concentrate on one at a time.

3. MAGIC TRIGGERS

I call special concrete objects "magic triggers" and use them as anti-anxiety devices to impart a sense of identity and calm in anxiety-provoking situations. When people feel anxious and cannot "think straight," simple objects, meaningful to them, can help. Usually in this society, we are so high and mighty that we think only ideas and thoughts will empower us. In fact, when anxiety throws us off, concrete objects are consoling in their solidity. "Magic" objects keep you awake and aware of what you set out to do and who you are. Like the good-luck charms that fairy-tale heroes carry on their journeys, special pieces of clothing and jewelry give power and a sense of identity. I suggest we enjoy our "magic" triggers to the fullest and infuse our experience of math with all the color, fun, flair, and extravagance we can muster. Since our culture is so poor in colorful, empowering rituals, we have to fashion our own.

This is not as outlandish as it may seem. I know powerful people in business and the arts who swear by lucky objects and numbers, or wear special rings or bracelets on trying days. There are two kinds of magic triggers, both of which remind us of our strengths, beauty, power, and intelligence. The first trigger should be with you at all times. The second trigger, a startling reminder for special events, should be worn or carried only for those events.

A TRIGGER TO CARRY ALWAYS

The character Sam, on the TV series "Cheers," always carries a bottle cap to keep him from drinking. Sam associates the cap with the last bottle of beer he ever drank, and it reinforces his resolution to stay sober. (In one episode he loses the beer-bottle cap and starts drinking.)

How would you choose a trigger to carry every day?

The object must be meaningful to you: perhaps the ring your grandmother gave you, or your grandfather's watch. This object keeps you in touch with your identity and reminds you of your purpose. There must be a connection between you and your "magic trigger." One of my students does his math with a "lucky pencil" he always carries with him.

You should carry your magic trigger so that it falls into your hands several times a day without planning it. I carry a crystal in my pocket, so that when I look for change it falls into my hands. The crystal conjures up my picture of the Alps, an image that reminds me of my roots and basic stability, strength, and beauty.

A TRIGGER FOR SPECIAL EVENTS

My student Dorothy always wore a heavy gold bracelet when she conducted department meetings or addressed groups of customers. The bracelet had been given to her by her mother, the mayor of a small city in California, who had always encouraged Dorothy in her career.

How would you choose a trigger for special occasions?

This trigger, like the one you always carry, works to remind you of power and the support of the universe, but for an event like an important presentation or meeting where you have to stand your ground, the magic object should be so *unusual* that you are reminded of your power every minute. Since it must startle you, obviously it isn't something you should wear all the time.

Sometimes people ask me if magic triggers are similar to teddy bears and security blankets. No. The magic trigger is less about comfort than about *leaving* your "comfort zone." Like the king whose heavy crown reminds him of his responsibilities, you are empowered.

4. SURROUNDINGS

Your environment is important for your thinking and learning—and so is the act of taking control of it. You don't have to be a slave to conventions like doing math on a lined legal-size yellow pad, or in pencil (because it's always "wrong" and you have to erase it), or in a quiet room, alone, or for at least two hours in a row. You can replace yellow legal-size pads with bright papers, and plain pencils with colored felt-tip pens. A quiet room or a library may work well for you, but many people find *silence* distracting, even isolating. Because I grew up in Switzerland, I like the tradition of working in cafés. Contrary to popular opinion, monastic surroundings are not a requirement for doing math. Lively surroundings, with the clatter of dishes and hissing of the cappuccino machine and the inimitable café conversation are comforting and make your math project seem more everyday and casual. Wear a beret for the French intellectual look and study Part IV ("Easymath") of this book, or prepare your department budget for Monday's deadline in your city's equivalent of a Greenwich Village bistro. Doing math can become a festive occasion.

5. LISTEN TO YOUR BODY

"I postpone work on the department budget and profit-and-loss statements until after dinner," says Paul. "But I always fall asleep. My wife says it's psychological fatigue, because she knows I hate doing it."

"Are you a morning person?" I ask.

"Yes."

"When do you get up?"

"Seven-thirty."

"Then why not get up at six-thirty and do your number-crunching first? Then come home from work and relax."

Lilly, an executive secretary in a marketing firm, and a part-time college student, takes math and a sandwich to a nearby park at lunchtime.

Mary, who knows that a "sea voyage" calms her nerves, reads stock market reports during her commute to Wall Street on the Staten Island Ferry.

For some reason, my students doubt at first that a change of rhythm or surroundings will make a difference in their success with math. In their anxiety, they have never taken charge of their day.

6. LISTEN TO MUSIC

Most people are trained to do mental work in quiet places, yet listening to baroque music has proven positive effects on thinking and feeling. Math learning and thinking in particular are closely related to music. If classical music is not to your taste, similar effects can be obtained from any music you like (it needs a good, regular beat, though). I know young people who do their hardest work with math and computers to the accompaniment of hard rock. *They* find it calming.

Instrumental music (passionate love songs might distract you) with a nice, steady beat works like a charm. Bach and Vivaldi are great choices. In *Superlearning* (Delta, 1979), authors Sheila and Nancy Ostrander and Lynn Schroeder describe the effectiveness of baroque music as a learning aid:

The idea of music as the bridge to inner awareness goes way back to the hidden sources of music itself. It runs deep in the legends of Orpheus who used music as a means of "charming" living creatures. . . . A few minutes a day of this baroque music, and listeners . . . began to report not only expanded awareness and better memory but also a whole repertoire of health benefits. They felt refreshed, energized, centered. Tension and stress disappeared. Headaches and pains went. The impersonal physiological graphs printed out proof—lowered blood pressure, lowered muscle tension, slower pulse. Is it just the *beat* of this music that slows body/mind rhythms to healthier levels or is there something else about this particular music that makes it appear to be especially life enhancing?

Try reading the rest of this book while listening to music, and use the technique when you work for any length of time with numbers. Your personal experience may bear out the research reported in *Superlearning*, that the rhythm and vibrations of music settle your body/mind/brain into a comfortable, focused and relaxed working mode.

7. REWARD YOURSELF FIRST

Rewards are wonderful supports. Most people believe rewards should come after a goal is reached. Yet, if you tend to be hard on yourself, if your anxiety gets out of hand, if you keep yourself on a tight schedule at all times, try a different tack.

Reward yourself first, do the work later.

Buying that theater ticket or new dress or spending an evening on the town, if you volunteer to handle your department statistics or before you perfect your accounting course, keeps that flagging motivation jumping back up to record highs.

This is the way it works: My friend Tim bought himself an expensive stereo set before a job interview that he knew included a math test. The stereo cheered and encouraged him every day while he prepared for the test. It made the interview more palatable, and took the punitive taste out of his mouth.

You might buy yourself that ''dress for success'' outfit *before* the salary review or *before* you undertake the training program to use the new spreadsheet. The early reward reminds you that you are taking the task seriously, that you are supporting yourself, that you absolutely deserve it.

8. MIND-DUMP EXERCISE

Free-flow writing is a wonderful anxiety-reduction exercise in which you time yourself (five minutes, fifteen minutes, even thirty minutes if you have time) and write absolutely anything that comes to mind.

Think of words triggered by the immediate situation: *boss, bills, presentation, percents, computer.* Put that word or words on top of a page. Then start writing furiously. Push yourself to write continuously, without lifting your pencil from the paper. You may not pause, go back, or read over what you have written. Your focus must be totally on speed: faster, faster. After a few minutes your handwriting tends to change, becoming more fluid, larger, sloppier. Let it. What you write can be ugly and obscene. You never have to read it!

I call this "dumping your mind," putting it all out with no criticism. No full sentences are necessary. This is not an essay. You have to write incredibly fast to catch your thoughts and outpace your reasonable mind, your surface mind.

The purpose of this exercise is to pull memories and marginal thoughts relating to the anxiety issue at hand out of the recesses of your mind. You have no time to think about or analyze your writing, but your mind will often make connections outside your awareness. You may suddenly feel much calmer and clearer because something clicked in the back of your mind, an association such as that every time you write a mind-dump, it is all about your father; or that all your associations have to do with being helpless and small; or that although you started with the cold, hard word *computer*, your writing veers off to descriptions of idyllic landscapes and past vacations—a sure sign that your anxiety is quite strong and that you are dreaming of the best possible escape.

DEALING WITH YOUR ANXIETY

Put the writing away. Don't read it immediately. You may want to throw it away or read it a few days later. If you are psychologically inclined, it may hold surprises for you.

9. KEEP A SUCCESS JOURNAL

Let your mind dwell on success. If you lose confidence, in your office, in the early gray morning of "billing day," do some math that you know well. I advise people to warm up with easy math problems before they deal with any number challenge. When you add up that column of numbers, you can believe again that your mind is working. Friends have told me that doing "mindless" math is a great tension breaker when they suffer over *The Wall Street Journal* or a spreadsheet. The Easymath sessions later in the book can be a great help in this, once you know them well. It does help your anxiety to do math problems over and over again. Every right answer registers as success, and deflates anxiety and self-doubt.

This is what your entries in the journal might look like. Noting your victorious moments in math counts a lot!

Jan. 15
I'm becoming a math genius! Counted change at stationery store and found it 20 cents short. Asked for correct change.

Jan. 10
Boss asked me to prorate vacation days for a secretary who's leaving. I made a drawing and worked it out. Boss pleased. I never thought I'd be able to do this.

Jan. 20
Blouse marked down to $18. The sign said 20% off. I knew the price was wrong. Wow! Look at me. I'm such a dandy smart person—at last.

Having your successes in black and white to read over and over helps your joy at your progress. It is also an interesting record of your journey to look back at later on.

Now that you're familiar with the eight Confidence-Building Techniques, you'll find that using the first three when the Math Monster strikes suddenly will alleviate anxiety. This is your Panic Kit: breathing, affirmations, and magic triggers. At such terrible moments when you're hit with blurred vision, nausea, headache and/or neck strain, you'll need to use these Panic Kit tools. They will help you stand your ground when the earth is falling beneath you.

Here's how Monique and James dealt with *PANIC*.

MONIQUE'S STORY

Monique was called on in a meeting to explain the 20-percent increase in departmental expenses over the last three months. She broke into a cold sweat. Twenty percent? She did not have any of the figures present, and she did not know how to figure percentages on the spot (or any other way, for that matter).
Panic struck.

HOW MONIQUE DEALT WITH HER PANIC

It was hardly the time for "mathematical confession." Monique remembered to slow her breathing and catch a glimpse of her affirmation written in bold letters across her daily agenda:

I am smart enough to handle numbers.

She found herself talking calmly, looking for the positive angle, and by ignoring momentarily the issue of the 20-percent increase, with her trusty "magic" silver pen she jotted down the changes that had taken place in her department: the expansion of staff, the new office machines. Hearing herself explain things helped. She estimated the cost of the new staff members and the machines. Holding the pen rather too tightly, she took a stab at estimating the new costs and stated that figure as her budget increase. Monique promised to have exact figures on her boss's desk by noon the next day.

The crisis was over. Nobody had pressed her on the issue of the 20-percent increase. She had spoken with authority, and her slow breathing had allowed her to think out her strategy clearly.

At the end of the meeting, Monique packed her silver pen in the special compartment of her briefcase and was glad she had defused her panic. That night she sat down with a friend (it's okay to get help!) and checked the departmental costs in detail and at leisure.

JAMES'S STORY

James walked into a large bookstore to find a book on Visicalc, a subject his boss had told him to learn.

James entered the textbook department and couldn't believe his eyes. Rows and rows of textbooks seemed to stare at him, and after a minute he felt them close in on him. Hundreds and hundreds of math books.

Panic struck.

He took flight and only felt better once he was half a mile away from the store.

HOW JAMES DEALT WITH HIS PANIC

After reviewing the contents of his Panic Kit, James went back to the bookstore and used his individual panic package:

1. Square breathing (remember: very slow to the count of four).
2. His favorite affirmation: *It's amazing what I can do.*

James had been writing his favorite affirmation on top of every sheet of notepaper at work, on the napkins at restaurants while waiting to be served. He listened to this affirmation mixed up with music on his Walkman.

James wore his beautiful huge pocket watch, which usually hung on the living room wall at his studio. It was a present from his grandfather, a token of admiration imbued with power, family appreciation, tradition. The watch made James feel appreciated and powerful.

James stood in the bookstore, staring at the rows of math books, holding on to his watch, saying in his head, *It's amazing what I can do.*

His tension eased and he calmly asked for the *simplest* visicalc book. He bought it, planning to read it at home, on his turf, and at a time that was comfortable.

You'll use techniques from your Panic Kit whenever numbers scare you at work or at home. You have more power than you think you do! If your boss confronts you in the corridor with questions about a major account or last month's sales figures, and you feel queasy (or worse), buy time. Be assertive. Agree to meet in a few minutes (or hours) in your office (or hers). That gives you time to use the panic techniques *and* to prepare your figures calmly.

VANESSA'S STORY

Remember Vanessa, whose math problems started as far back as she could remember? Vanessa, you recall, shrank at the sight of a math problem, avoided math through school and college, but needed to get a high score on the Graduate Record Exam to get a master's degree in clinical psychology. I agreed to tutor her twice a week at the Hungarian café. She liked the distractions and being able to flee, if only to the counter to order another coffee. When things really got bad, she would get up and order pastry.

The best cure for math anxiety is instant, steady success, in small doses and at all times. I built Vanessa's math program to ensure her success. It took many positive experiences for her to start counteracting all her negative images as she learned basic computations and I showed her every intuitive and common-sense way of solving problems—trial and error, estimating, drawing—all strategies you'll learn in Part IV.

Vanessa learned arithmetic, basic algebra, and basic geometry at a reasonable pace. Suddenly she hit a plateau. Losing confidence for the first time, she was willing to give up. "I guess I just can't learn math," she said.

It was time to use all the Confidence-Building Techniques. I knew that Vanessa studied T'ai Chi, a martial art that develops balance, coordination, breathing, and stamina. I believed that if she worked with her breathing and posture, her anxiety would lessen and her math would improve. We rescheduled the tutoring hour so that she could come right after her T'ai Chi class, and were able to work steadily for longer periods. Vanessa was willing to try affirmations. She liked *I am brilliant*, listened to it on her tape recorder at night, and scribbled it on top of her math worksheets. As the weeks went on, Vanessa's voice took on a deeper, more assertive tone. Her stance was more open. She smiled more easily. (A real warrior woman!)

Merrily chugging along, Vanessa started to take pleasure in some of her math. "I really like geometry," she said one day. Only a few weeks earlier, this had seemed impossible.

When the GRE approached ("I can't sit for so long. I won't be able to breathe."), we worked out strategies. Vanessa scheduled a T'ai Chi class and a quick math warmup the day before the test and took along to the test a cherry-flavored lollipop and a colored pen for doodling in the margins. She asked to borrow my large batik scarf to serve as a trigger, prompting her affirmation in an inner voice, calm, steady, helping her along.

I suggested she wear her T'ai Chi outfit to take the test. Then, every fifteen minutes (Vanessa's natural working interval), she could get up and do a short T'ai Chi sequence. We both laughed, thinking the proctor might be thrown for a loop as he spotted the Karate Kid amid a sea of regular students. We agreed she would arrive at the test early to secure the corner back seat where she wouldn't disturb anyone. Do you see how much assertiveness this took?

Vanessa aced the GRE. All the Confidence-Building Techniques worked; the tutoring in the comfort and hubbub of the Hungarian café, the T'ai Chi, the scarf, the affirmations, the lollipop—every technique helped her stay clear and focused. Most of her new math skills worked, leaving her with an impressive score of 610 and a passport to a good graduate school. In the end, it was Vanessa's strong persistence more than anything else that determined the successful outcome. She was perfectly capable of handling any math expected of her, including the infamous graduate-school bugaboo: advanced statistics.

The time has come.

You're ready to go on to Easymath. You can start the next section even if you think you're not quite ready. Learning math and reducing your anxiety go hand in hand. And don't get rid of *all* anxiety. Mild anxiety is a benign state, full of excitement and promise. Actors know that a little bit of stage fright works for a good performance. Just a little anxiety keeps you revved up. After all, if your life were totally free of stress, you might just sit there like a lump.

Once you get your feet wet, simple arithmetic is an anti-anxiety endeavor. I have both math-anxious and math-loving friends who find arithmetic—adding, multiplying whole numbers—reassuring and restful. Numbers, uncomplex and reliable friends, are steady and don't turn on you overnight. You may even find working with them fun!

IV...
EASYMATH:
JUMPING OFF THE CLIFF

▪ ▪ ▪
STARTING OUT

Jumping off the cliff is an energetic metaphor. Get into it, heart and soul. Risk everything. That's what we have to do when we want to move on in life. How would it help to see learning math as such an earthshaking undertaking? Ask yourself how many times you have put it off, saying you had no time for this class or that workshop, or have forgotten to pick up a book? We all may need a push once in a while to go in a new direction, but it is more fun to take a leap than to be shoved. So be in charge of your own adventure.

Learning is not that dangerous. Sometimes I think that is unfortunate. Maybe school and education and learning would be a lot more appealing if they were a little more extravagant and dangerous. But you can make them so. After all, you're the master of your fate. You can add adventure by placing bets on when you'll finish working through this book, or you can get together with a group of gung-ho people and do the math that follows together, or you can read it while whitewater rafting, or you can promise yourself a giant reward. It is your life and your adventure.

Although the name Easymath may sound like a self-contradiction, I'll prove that you can learn the crucial skills within a short time.

Handling numbers in daily business has to be simple because it was invented by regular people to make things *easy.* (No, math was not invented to torture little children!) *Percentages* are a most important topic, but they give trouble to a lot of people.

Simple fractions were easy for you when you were a little

child; half a cookie or a quarter of a pie is easy for a child to figure out, and children today most certainly know which TV program starts on the hour or the half hour. If Adam and Eve had thought about fractions and split their apple fairly down the middle, into equal halves, the whole war of the sexes might have been avoided. Lest you think this is too heavy a topic to bring up in this context, I can assure you that many of my students have found fractions painful on an emotional level. Janet, a secretary, actually got ill at the sight of one of those ugly little beasts, ¾ or ⅝. Delving into her life history, we found out that her parents divorced while she was in the fifth grade, the very time her math teacher asked her to divide nice, wholesome circles into four or eight equal pieces. I have found this scenario quite often. Our deeper mind does not like division and fractions if it has reason to connect them with painful memories of childhood. (Who gets the house? Who gets the kids? Who gets the dog?) Once Janet was aware of what her "fraction terror" was all about, she could *separate the issues* and give the numbers back their simple use and purpose.

Estimating is a skill that experienced math people use constantly, and it is this skill that makes them seem so mysteriously fast and competent. Handling figures quickly in your head can bring positive results on the job and in other aspects of daily life.

Rita came to me overwhelmed from a budget meeting where she had been stunned at her colleagues' speed at figuring money allocations for various departments, working only from an intimidating-looking computer output. What would *you* do with 9.5% of $20,975?

The trick, which seemed too bold and inexact to Rita, is to round off the numbers dramatically and rewrite them in your head as: 10% of $20,000, at which point it becomes an easy problem. (The answer is $2,000, but more about that later.)

Most schools have swallowed whole the myth that math is

always about exact answers—an incorrect and intimidating concept that hinders people in real life.

A LITTLE MATH QUIZ

Answer these questions any way you can—by guessing, estimating in your head, on paper, with a calculator, by divine inspiration. Remember, this is for your eyes only. Have fun. It doesn't matter how many questions you can answer; we'll learn it all in the next chapters anyway.

20 QUESTIONS

1. Tip: 15% on $24.
2. $\frac{1}{2} + \frac{1}{4} + \frac{1}{8}$
3. Take 30% off a price tag of $800.
4. Movie: 2 hours 11 minutes. How many minutes?
5. Drive at 55 mph for six hours. How far?
6. Half of $\frac{1}{10}$?
7. Temperature 22 degrees F. at noon, drops 30 degrees by midnight. Temperature at midnight?
8. 4.5×4.5
9. Six and a half dozen pencils. How many?
10. A new airplane costs one-half billion dollars. Write it out.
11. One square yard is how many square feet?
12. Flip two coins. Chance for two heads? (Try it.)
13. $28 + 29 + 30 + 31 + 32$?
14. Half of $3\frac{1}{2}$ hours?
15. Houses on the right side of a street are numbered from 12 to 172 (even). How many houses?
16. Fold a paper in half four times. How many layers in the wad? (Try it.)
17. How many quarters in $7?
18. Three-quarters of a pound equals how many ounces?

19. Wholesale cost of a dress is $50. Selling price is $200. Markup?
20. A suit costs $100. The jacket costs $1 more than the skirt. How much does the jacket cost?

I hope you looked at each problem. More than once. Trying is essential, even if you're not optimistic and even, in fact, if you're quite sure you have no idea what you're doing. Trying is good because it makes you feel in charge and keeps you going. Nobody is grading your quiz; I hope you won't, either. Checking the correct answers should just be fun. You want to know where you were right and where you were wrong. By the way, don't throw away the paper. It's going to be interesting later. As you go on, you can go back and take the quiz again, and you'll do better every time.

ANSWERS

1. $3.60	11. 9 square feet
2. $7/8$	12. $1/4$ or 25%
3. $560	13. 150
4. 131 minutes	14. 105 minutes
5. 330 miles	15. 81 houses
6. $1/20$	16. 16 layers
7. −8 degrees	17. 28 quarters
8. 20.25	18. 12 ounces
9. 78 pencils	19. 300%
10. 500,000,000	20. $50.50

How did you do? What did the quiz remind you of? Did you hate having a quiz slapped in front of you when you were a kid in school? Did this feel the same or different? If you got all of them wrong, this is the book for you, and I'm glad you've decided to start dealing with your math.

If you skipped the quiz altogether, that's all right, too. You may not be ready to look at many math problems all on one page. Take a break and look over Part III for some comforting techniques, practice them, and come back to Easymath. You'll be fine.

SESSION 1.

PATTERNS

"I should not be allowed in here. I don't even know my multiplication tables," one of my students says at the beginning of a workshop. Do you feel like that? Yet I am sure you know some of them: the twos, the fives, the tens. Which multiplication tables do you like and which do you loathe?

You may consider this a frivolous or superfluous question, but it is as important in math as in other areas of life to be clear about your predilections, because our feelings influence our thinking and our motivation.

Take a second to notice what the easiest tables (twos, fives, tens) have in common. Two, five, and ten all go into ten, and our whole number system is based on ten.

Usually people don't mind threes, fours, and sixes very much, but they often dislike sevens, eights, and nines.

There is no quick help for the sevens! Seven is an oddball in our number system; it does not have a pleasant relationship with any of the powers of 10: 10, 100, 1,000, etc.

Nine, however, is smooth and elegant, sitting as smug as can be right next to 10. So, by the way, is 11. Look at these two lists for a moment. Look for patterns!

$1 \times 9 = 9$	$1 \times 11 = 11$
$2 \times 9 = 18$	$2 \times 11 = 22$
$3 \times 9 = 27$	$3 \times 11 = 33$
$4 \times 9 = 36$	$4 \times 11 = 44$
$5 \times 9 = 45$	$5 \times 11 = 55$
$6 \times 9 = 54$	$6 \times 11 = 66$
$7 \times 9 = 63$	$7 \times 11 = 77$
$8 \times 9 = 72$	$8 \times 11 = 88$
$9 \times 9 = 81$	$9 \times 11 = 99$
$10 \times 9 = 90$	$10 \times 11 = 110$

The elevens show nicely how both digits go up 1 (we add a 10 and a 1). Do you see a pattern with the nines? Vertical? Horizontal? Do you see that the first digit goes up 1 and the second digit goes down 1? Does it make sense?

SEEING PATTERNS MAKES
MATH INTERESTING AND SAVES TIME

The number 9 can be most easily worked with if we realize it's just 1 short of 10, so adding 9 quickly in your head, you can add a 10, take away a 1:

$$57 \quad \rightarrow \quad 67 \quad \rightarrow \quad 66$$

add 10   take away 1

$$1{,}024 \rightarrow 1{,}034 \quad \rightarrow \quad 1{,}033$$

add 10   take away 1

Multiplying by 9 quickly in your head, you go:

$$10 \times \text{number} - \text{number:}$$
$$9 \times 45 = 10 \times 45 \text{ take away } 45 = 450 - 45 = 405$$

Do: 9 × 35 = ____ 9 × 12 = ____ 9 × 18 = ____
Answers: 315; 108; 162.

The same method works for 99, which is 1 short of 100: (100 × number − number):

   99 × 5 = 100 × 5 take away 5 = 500 − 5 = 495

   Do: 6 × 99 = 6 × 100 take away 6 = 600 − 6 = 594

Note that in the above, as elsewhere, we make two easy steps in place of one harder one. In mental, quick arithmetic, it is good to break things down into small, easy steps.

Checking for nearby easy numbers (999 is one short of 1,000, 201 is one more than 200) is an excellent way of looking at numbers to work in your head.

When you work with numbers, think of real-life equivalents as often as possible:

25 → quarters   8 × 25 (8 quarters) = 200

60 → 1 hour = 60 minutes   ¼ of 60 (quarter hour) = 15

100 → $1 (100 cents)   ¾ of 100 (3 quarters) = 75

QUICK TRICKS

MULTIPLICATION

To multiply a whole number by 10, 100, or 1,000, attach one, two, or three zeroes respectively.

To multiply a number by 5, you cut it in half and then multiply by 10.

EXAMPLE:

$$288 \times 5 \text{ is } 144 \times 10 = 1,440$$

That is painless, isn't it?

DIVISION

Since division is the reverse operation to multiplication, all the rules work in reverse.

To divide a number by 10, 100, or 1,000, take off one, two, or three zeroes respectively. (If the number has no zeros, just cut off the appropriate number of decimal positions.)

EXAMPLES:

$$45,000 \div 10 = 4,500$$
$$367.9 \div 100 = 3.679$$

Practice:

$7 \times 10 = \underline{\hspace{1cm}}$	$5 \times 100 = \underline{\hspace{1cm}}$
$73 \times 10 = \underline{\hspace{1cm}}$	$52 \times 100 = \underline{\hspace{1cm}}$
$730 \times 10 = \underline{\hspace{1cm}}$	$100 \times 100 = \underline{\hspace{1cm}}$

$$9 \times 1,000 = \underline{\hspace{1cm}}$$
$$97 \times 1,000 = \underline{\hspace{1cm}}$$
$$253 \times 1,000 = \underline{\hspace{1cm}}$$
$$1,000 \times 1,000 = \underline{\hspace{1cm}}$$

SEQUENCES

Just relax and see if you can continue the following sequences. Can you guess my rule?

a. 1, 4, 7, 10, 13, ____ , ____ , ____

b. 1, 2, 4, 8, 16, ____ , ____ , ____

c. 1, 3, 7, 15, 31, ____ , ____ , ____ , ____

d. 128, 64, 32, 16, 8, ____ , ____ , ____

In (a) you might notice first that the sequence goes up; it grows nicely at equal intervals, up 3, up 3. The answer is 16, 19, 22.

In (b) the sequence is going up too, but it's growing "irregularly"; look again, times 2, times 2, times 2. The answer is 32, 64, 128.

In (c) things are less clear. It grows, but how? Up 2, up 4, up 8, up 16; that's not too bad.

The next jump is 32, so the next number is 63, then 127, then 255.

By the way, this is not the way I had thought of sequence (c) originally. Look at this sequence again; it can be seen totally differently. My pattern is double and add 1, double and add 1, etc. Check it, it works:

$$1 \longrightarrow 3 \longrightarrow 7 \longrightarrow 15 \rightarrow 31 \rightarrow 63 \rightarrow 127 \rightarrow 255$$

double + 1, double + 1, double + 1, etc.

$$2 \times 15 + 1 = 31$$
$$2 \times 31 + 1 = 63$$
$$2 \times 63 + 1 = 127$$
$$2 \times 127 + 1 = 255$$

Why worry about sequences? Sequences are a quick and pretty way to think about numbers. They show up in nature and in business (monitoring developments, making predictions, figuring the half-life of radioactive substances, your income over years, the weight gain of your baby).

To solve problems from discounts to pay raises to the probability of rain, you need to look at numbers in relation to each other. You need to ask questions like these: Is it bigger, smaller? Is it half or double? Is it 10% more or 10% less?

We do sequences to show relationships.

SESSION 2.
WHOLE NUMBERS

You learn math the same way you learn to walk. You try, fall down, get up. The mere fact that you're reading this book means you have let go of the railing or of your mother's hand, and you're sailing straight ahead. If you're scared, if you think you're stupid or that math does not fit into your skull, so be it. Just turn the page and keep going. Math is not a mental enterprise only. It is hard physical work. Learning is physical. You write, you lean forward, you rewrite. You talk to yourself. You go back and do it over. Math is a lot of physical activity. Learning math is not like having chicken soup ladled into your mouth; it's like going hunting for your food. Cave dwellers got hungry first, and then went hunting. No question about motivation!

That's how learning is when it's good. You want to know something so much you can taste it. You must sit at that table and draw and write and scribble and go on and understand that it does not fall into place in the first run-through.

ONE-PAGE-A-DAY MATH

Let's make a contract right now. If this math is new for you, you must promise me you will do one piece a day. One paragraph. One page. One double page. Decide on a basic portion. No more, no less. Everybody understands the reason for doing no less; after all, we want to get through the book. The reason for doing no more is just as important. It will be hard to stop at the end of your page. ("Just as the going got good . . ."; "I was just getting into the math mode"; "I have so much time on

Thursday, maybe I should cover seven pages today and then skip three days.") See, that's where the chaos starts, the compensating. It's like going on a diet. Instead of eating like a normal human being, one says, "I had a big dinner yesterday, so I can't have lunch today"; "I had starch in the morning, so I can't have any for dinner." Your mind starts figuring everything except the work at hand. Okay. One page a day. You won't want to stop, and you'll be eager to go back to it the next day.

If you're just reviewing this material, you'll obviously concentrate on the pages that you most need and/or least remember.

MENTAL ARITHMETIC

Mental arithmetic is a lost art in this country, along with memorizing songs and poetry. Doing mental arithmetic and memorizing poems can be different from rote learning if you go through the process every time; you don't just recall dead facts from memory. Being able to figure in your head brings a sense of strength and competence. When you can figure out things in your head, you feel good, you look good.

I walked into the classroom and said, "Let's do mental computations today." Several students panicked, and I asked them what was so terrible about it. One woman said, "I always lose places, or the factor or the divider, I lose something," and another woman said, "I can never do it fast enough."

So I gave them simple problems to do. For example: *5 times 5. Double it. How many fives in that? Divide that by 4. How many quarters in that?* And the answer should be 10. It is important to know that for mental arithmetic you need totally different methods from those you use for written arithmetic. I find people trying to visualize a piece of paper or a blackboard, which is fine, but they're doing the problem with carrying and borrowing and everything the way they would do it on a piece of paper.

That's too complicated. I couldn't do it like that. I could not keep up my concentration and see all these little numbers. In mental arithmetic you have to break every operation down into small steps. A mind can handle many small steps a lot more easily than one big step.

USE THAT MIND OF YOURS!

Let's start. I'll give you a chain problem. Don't write. Just figure in your head, if you can. It will help keep you young and alert. (A chain problem uses the answer to one problem to lead to the next, and so on.)

EXAMPLE: $4 \times 25, \div 2, \times 5, \div 10, \times 8$

Step 1. $4 \times 25 = 100$
Step 2. $\quad\quad\quad 100 \div 2 = 50$
Step 3. $\quad\quad\quad\quad\quad 50 \times 5 = 250$
Step 4. $\quad\quad\quad\quad\quad\quad\quad 250 \div 10 = 25$
Step 5. $\quad\quad\quad\quad\quad\quad\quad\quad\quad 25 \times 8 = 200$

Half the challenge is the arithmetic, and half of it is keeping the numbers in your mind. Here are three of them for you to do!

1. 4×25	2. 6×6	3. $200 \div 4$
$\div 10$	$\div 3$	times itself
$\times 18$	take half	$\div 100$
$\div 4$	$\times 25$	$\times 8$
$\div 5$	$\div 10$	$- 101$
$\times$ itself	$\times 4$	$\div 9$
$+ 19$		

The answers you got, if correct, are: (1) 100; (2) 60; (3) 11. Let's look at some of the steps:

1. When multiplying by 25, think money. Think quarters. See quarters on the kitchen table. How would you count them, by fours? $4 \times 25 = 100$, $8 \times 25 = 200$, $12 \times 25 = 300$, etc.

2. Multiply or divide by 10 or 100 according to "Quick Tricks" on page 90.

3. Multiplying or dividing by 4 comes up a lot in real life: quarterly payments, 25%, etc. Rather than do multiplication or division by 4 in your head, we think better if we do it the way we would in practical situations. How do you divide an apple pie four ways? Cut in half. Cut in half again.

Do that with numbers: $200 \div 4$. Half of 200 is 100; half again is 50.

By the same token, multiplication by 4 can be done in two easy steps: Multiply by 4; double and double again. Multiply 15×4: 15 doubled is 30; 30 doubled is 60.

4. Subtract 101. Do it in two steps: Subtract 100. Then subtract 1. For example: $200 - 100 = 100$; $100 - 1 = 99$.

Remember: Your mind does better with many small steps than with one long, complex procedure.

People who are experienced in math tend to play with a problem before solving it. Let's look at a multiplication like 8×45. This can be simplified quickly:

$$\div 2 \left(\begin{array}{cc} 8 & \times \ 45 \\ \\ 4 & \times \ 90 \end{array} \right) \times 2$$

$$\div 2 \left(\begin{array}{cc} 4 & \times \ 90 \\ \\ 2 & \times \ 180 \end{array} \right) \times 2$$

Do you see that the above three ways of expressing the problem all come out to the same thing (i.e., 360)? Many steps, no trouble, no pain. Try it:

$$\div 2 \left(\begin{array}{cc} 16 & \times \ 55 \\ \\ 8 & \times \ 110 \end{array} \right) \times 2 \qquad = 880$$

$$\div 2 \left(\begin{array}{cc} 18 & \times \ 4\frac{1}{2} \\ \\ 9 & \times \ 9 \end{array} \right) \times 2 \qquad = 81$$

OUR NUMBER SYSTEM

Our mathematical language isn't clear to many people. I have found out that some people truly have no idea that attaching a zero to a number makes it ten times bigger. They see absolutely no way of explaining, when two numbers are added together, what the "carry" is all about. They can't understand why we start adding with the rightmost position, and what the carries are.

To appreciate our number system, you have to imagine others. Do you know Roman numerals, like the ones on historic buildings?

$$CXXIII = 123$$
$$CXC = 190$$
$$MCCCIII = 1,303$$
$$MMX = 2,010$$

Notice that Roman numerals just enumerate the *ingredients* of the number; the length of the number bears no relation to its size. Would you like to do long division with those entities? I know you wouldn't.

Now let's look at the way we have come to write down our numbers. Our system is of Indian-Arabic origin and has the incredible advantage of lining things up in columns. For example: 245 means two hundred plus forty plus five; 3,097 means three thousand plus no hundreds plus ninety plus seven.

The best thing to do at this moment is to run out and get some graph paper. Math competency and test scores would zoom up in this country if we did our arithmetic on graph paper. After all, accountants use vertical lines to guide their figures correctly. So let's not quibble with it.

hundreds	tens	ones
3	2	5

thousands	hundreds	tens	ones
3	0	2	5

Notice that when we write our numbers, we make a lot of assumptions. We slot our digits into "invisible" slots and we all understand what we mean. There is a big difference between 325 and 3,025. Although zero in itself has no value, here it is a place-holder. Draw in the columns in your mind or on your notepaper. The first number reads three hundred twenty-five, the second number reads three thousand twenty-five. The zero tells us there are no hundreds.

WHAT DOES "CARRY" MEAN?

In my workshops, I put an addition question on the board like:

$$\begin{array}{r} 175 \\ +239 \\ \hline \end{array}$$

Everybody does it, but nobody can quite explain why we start at the back of the number, and what the "carry" is. So I go up to the board and play dumb, and my students dictate to me: "Nine and five are four; put down the four and carry the one." I answer, "What am I carrying, and where to, and why a one?"

People almost don't *expect* to understand. I tell you, just put in the columns (that is the secret behind all whole-number arithmetic!).

hundreds	tens	ones		hundreds	tens	ones
1	7	5		1	7	5
+ 2	3	9		+ 2₁	3₁	9
3	10	14		4	1	4

with carrying becomes

The left part of the above figure is a naive way of adding correctly, except there is room for only *one* digit in each column! So 14 has to be split up into four ones and one ten, and carrying the ten over to the next column is what the "carry" is all about.

WHAT IS "BORROWING"?

Most people remember vaguely about borrowing. Borrowing is needed in a subtraction like 1,000 − 432, e.g., "Is it all nines on top?" they ask. As usual, it is best to be practical. Don't try to remember way back to the sixth grade. Try to act it out. Say

I have a $1,000 bill in my house and I have to pay somebody $432. I just cannot do it. I have to break the big bill. I go to the bank and get ten $100 bills. Then I break one of the $100 bills into tens and I break one $10 bill into singles. Now I have on my table a heap of bills: nine $100 bills, nine $10 bills, and ten $1 bills. That is what borrowing is all about!

Breaking the digits up into columns, we have:

$$\begin{array}{c} 1\mid 0\mid 0\mid 0 \\ -\ \ \ 4\mid 3\mid 2 \end{array} \quad \text{which may now be rewritten as} \quad \begin{array}{ccc} 9 & 9 & 10 \\ \cancel{10} & \cancel{0} & \cancel{0} \\ -\ \ 4 & 3 & 2 \\ \hline 5 & 6 & 8 \end{array}$$

Now the transaction goes smoothly.

THE TALE OF THE LITTLE SNAIL

Are your eyes beginning to glaze over from all the digits and columns? Then it's a good time to let go and change pace. We need a good puzzle. Arithmetic is straightforward; puzzles are not. Puzzles ask for common sense and a good dose of right-brain thinking (seeing things in context).

Puzzles are problems with a twist! This puzzle, which I worked out with my client Freddy, is about a little snail, and is straightforward as puzzles go.

"There's a little snail and a flagpole, and the flagpole is 11 yards high," I explain to Freddy. "This is crucial information. The little snail has an incredibly ambitious plan. He wants to climb the flagpole. He is the first in his family ever to attempt that feat. Most of his relatives stay close to the ground. Now he is anxious and nervous. We know about that. When you attempt something for the first time for everybody in your family, you are going to be nervous. So, on the first day, he gets up, the sun is shining, he feels great, he climbs up 3 yards. Night falls, it gets dark, it gets lonely, it gets cold, he gets the shivers, he slides 2 yards down. Notice, only 2 yards down. The next morning the birds are singing, life is good, his courage returns, he goes up, he does the same thing, he goes up 3. At night he goes down 2, and he keeps up this pattern until he hits the top. If we say he started on Day 1, on which day—and don't answer in two seconds—on which day does he hit the top? It's like a space mission. We call it Day 1, Day 2, and so forth.''

Freddy says, ''The answer is Day 11.''

''Yes, that's the logical answer. Very logical left-brain thinking. Every day the snail goes up 3, goes back 2, and makes 1 yard a day. The flagpole is 11 yards high, so it takes 11 days. That's logical, but it has nothing to do with the problem. Now I'm going to give you a hint. First, draw on lined paper or graph paper a flagpole exactly 11 spaces high. Then, imagine you are that little snail. Show on paper what he does—up, down, up down. Don't think, just draw what he does. Can you see that as he gets toward the top, the zigzag pattern is interrupted?

Freddy says, ''No, I don't.''

''Look at the picture,'' I tell him. ''On Day 8 the snail spends the night at altitude 8 yards, and then what happens? Act out what the snail does on Day 9.''

Freddy says, ''He climbs 3 and comes down 2 that day as usual.''

''No, no,'' I say. ''He gets up at altitude 8 yards, goes up 3 yards, and now it happens. He reaches the top and plants his little flag. He's done it.''

Freddy says, ''Well, I see that, but I see it my way, too.''

''But your way doesn't happen. When you said 11 days, you made the snail overshoot the top. Look at it again. On the ninth day he goes up his usual 3 yards and he's at the top, he's there. You forced him to come down 2. The question was, on which day does he reach the top? And that's on Day 9.''

I believe word problems and puzzles have to be acted out. They don't have to be thought out. That's why it's important that you imagine the animal, what is happening, what it is doing, up, down. Quick left-brain reasoning is often false. There are very often what are called marginal issues. Things are very often different at the end—at the end of a year, at the end of a trip, at the end of something. The more concrete you are about these things, the better.

SESSION 3.

PAINLESS PERCENTS

I think math is a mystery to most people. It is often not taught well. We would be better off if teachers just chose to teach math only when they loved it, and left the rest to the students, since most math in daily life needs few sophisticated skills and plenty of common sense. What our schools do is make math seem remote and complicated, and so schools effectively take math away from the students, making them distrust their own perceptions and thoughts, depriving them of their innate, natural tools to solve mathematical problems.

How do we find our way back to original thought? How do we dig out our sense of fun with numbers and shapes that was so obvious when we were little? By playing. Can you still play? Checkers or chess? Pool or tennis? There is math in all of them. Maybe you know about crafts. Do you do carpentry? Wallpapering? Sewing? Tiling? Cooking? Weaving? I suggest that you look at all the things you do with a new eye, looking for rules or patterns or regularities. I myself don't drive, but I'm sure that parking, passing, and weaving from one lane to the other have plenty to do with angles and knowing about distance, speed, and time.

Let's start every session with chain problems!

MENTAL ARITHMETIC

	1. ½ of 600	2. ½ × 1,000	3. ¾ × 100
	÷ 6	÷ 4	÷ 5
	× itself	÷ 5	× 100
	÷ 100	× 8	÷ 3
	× 5	× ¼	× ½

The answers are (1) 125; (2) 50; (3) 250.

PAINLESS PERCENTS

What are percents? A very basic question, but this is the only country where people cannot answer it clearly. "Parts of a whole," they say, looking vague and uncomfortable.

"Let's take 15 percent; what part is that?" I ask.

Often only one or two people in the workshop know.

"Any linguists here? Anybody here who speaks French, Spanish, Latin? What does the word mean?"

Now you will learn percents once and for all in five minutes. Here goes. If you speak any Spanish, French, Latin, or Romanian, you can translate *percent* into plain English as "per hundred." *One percent* means "one per every hundred." If you think about it in terms of money, it means one cent on every dollar. See?   1 " out of " 100

TIPPING

In most parts of the world, the standard tip is 15%. That is simply fifteen cents on every dollar. If you have lunch for $5, that means 5 × 15 cents = *75 cents*.

We did it!

We just learned to tip at a restaurant. Trivial, you think?

Pause for a moment. Most of business math, interest, pay raises, markups, discounts, taxes, all run on this principle.

"Yes, but what if the numbers are harder?"

"Yes, but what if I don't understand business?"

Were you thinking that way? Math-anxious people are always a little impatient, impatient with simple things, impatient for the tough stuff. Yet the simple things, *fully understood,* will serve you best.

If you know how to tip, you can figure:

> *Interest* of 9% = 9 cents on the dollar
> *Discount* of 20% = 20 cents on the dollar
> *Pay raise* of 4% = 4 cents on the dollar
> *Growth* of 8% = 8 cents on the dollar

How do you feel now?

Paul looks skeptical. "You're just trying to make us feel good. I could never figure interest, I don't understand banking!"

Sonia says, "I always have to leave as soon as two or three workers get together. They discuss union business, 17 percent raises over three years, and so on. I have no idea, so I leave."

Let's help Paul and Sonia. Percents are a very simple concept. They have to be, for the system most probably was developed by ordinary people like you and me. Maybe a fisherman in Italy was down on his luck and needed to borrow money for his food and Chianti. The local moneyed person charged for the favor by saying, "Give me back 5 lire extra for every 100 you borrowed."

That's it. An everyday concept. Let's figure the interest on the fisherman's loan.

LOAN	PERCENTAGE	ANSWER
600 lire	5%	5 lire on every 100 makes 30 lire

SIMPLE ATTACK METHODS
FOR PERCENTS

The method we have used so far always works. It is the most basic way of doing percents. We'll call it Method 1.

Yet sometimes there are more suitable, quicker ways to do things. Method 2, computing 10% first, provides a stepping-stone and is helpful in estimating and in figuring round percentages. Method 3, computing 1% first, is helpful for the more involved percentage problems, as you will see from examples later. All three methods come up with the same answer, of course; you just choose the one that you like best for each situation. Let's make sure:

EXAMPLE:

A coat costs $400. You buy it during a 20% sale.

Method 1. 20% discount means you get 20 cents off on every dollar. That is: 400 × 20 cents = 8,000 cents = $80.

Method 2. Compute 10% first:
10% of $400 is $40.
20% is twice as much, or $80.

Method 3. Compute 1% first:
1% of $400 is $4.
20% is 20 × more than 1% = 20 × $4 = $80.

THE 10% METHOD

Method 2 (computing 10%) is the method of choice for slick, quick math. It's the best way to hold your own at a budget meet-

ing, at a family money conference, in the store—in short, anywhere you don't want to be caught doing lengthy computations
and you need a quick insight into what your share is. Tips,
discounts, estimates—there are dozens of situations that can be
handled by "round percents": 10%, 20%, 30%, etc., or 5%,
15%, 25%, etc.

Since we live in a number system totally built on 10 (the
decimal system), taking 10% is easy. Taking 10% on every 100
means a dime on the dollar. It's the same as dividing by ten.

10% of $100 is $10	10% of $60 is $6
10% of $400 is $40	10% of $7 is $.70
10% of $250 is $25	10% of $5.40 is $.54

Just notice what happens in each of the above examples. The
decimal point jumps one position to the left. Finding 10% is an
easy first step before going to other "round" percentages.

COMPUTING ROUND PERCENTAGES

1. 30% of $400.
Do: 10% of $400 is $40
    30% of $400 is 3 × $40 = $120

2. 70% of $500
Do: 10% of $500 is $50
    70% of $500 is 7 × $50 = $350

3. 5% of $20,000
Do: 10% of $20,000 is $2,000
    5% of $20,000 is $1,000

4. 45% of $300
Do: 10% of $300 is $30
    40% of $300 is $120
    5% of $300 is $15
    45% = 40% + 5% = $120 + $15 = $135

USING THE 10% METHOD FOR ESTIMATING

1. Interest: 9.7% of $2,387.79
Estimate: 10% of $2,400 is $240.
Keying sequence on calculator if you want to check:
2387.79 $\boxed{\times}$ 9.7 $\boxed{\%}$
Exact answer: $231.62

2. Pay raise: 4.8% on $30,907.79
Estimate: 5% of $30,000
Do: 10% of $30,000 is $3,000
   5% is half of 10% is $1,500
Exact answer: $1,483.57

Do you see how good the estimates are? In most real-life situations (budgeting, estimating costs and discounts), the precision of the estimates is more than satisfactory.

3. New York City sales tax (8.25%) on $124.95
Estimate: 10% of $120 is $12
Exact answer: $10.32

For planning purposes, for quick comparisons, for finding gross errors, the estimate is good enough. And you can do it in your head.

QUICKIES FOR PRACTICE

Hint: Use 10% as a stepping-stone. Make use of previous answers if you can.

	ANSWERS
20% of $100	$20
20% of $300	$60
20% of $50	$10
20% of $250	$50
20% of $12.50	$2.50
5% of $100	$5.00
5% of $300	$15.00
5% of $50	$2.50
5% of $250	$12.50
5% of $1,250	$62.50
5% of $12.50	$.63
15% of $100	$15.00
15% of $300	$45.00
15% of $50	$7.50
15% of $250	$37.50
15% of $1,250	$187.50
15% of $12.50	$1.88

Not so quick:

	ANSWERS	
10% of _____ is $450	$4,500	$450 = 10% of ?
20% of _____ is $10	$50	$5 = 10% of ?
30% of _____ is $75	$250	$25 = 10% of ?

Every percent problem can be solved with Method 3 (figuring 1%). This is a sure-fire method, and it always relieves anxiety to have one sure-fire method. It is not always the quickest way, but it always works and is essential for problems where we are looking for the interest rate or the capital.

Remember the methods we have learned so far:

EXAMPLE:

Lunch tab $6; tip 15%.
Method 1. 15% is 15 cents on the dollar
            6×15 cents = 90 cents

Method 2. 10% of $6 is 60 cents
           5% of $6 is 30 cents
           15% of $6.00 is 90 cents

Method 3. 1% of $6 is 6 cents
           15% of $6 is 15×6 cents = 90 cents

Remember:

$$1\% = 1 \text{ cent on the dollar}$$
$$1\% = 1 \text{ per hundred}$$
$$1\% = 1/100$$

Taking 1% means dividing by 100.

$$1\% \text{ of } \$100 \quad = \$1$$
$$1\% \text{ of } \$20{,}000 = \$200$$
$$1\% \text{ of } \$3.00 \quad = \$.03$$

Note: If the number has zeros, just take two off. Otherwise move the decimal point two positions to the left.

There are only three essentially different percent problems. So far we have consistently figured *percentage*—discounts, pay raises, etc.—which answers the easiest and most frequently asked questions about percents. Sometimes, however, we need to know other parts of the story, such as the capital or the interest rate. Since percents describe such a simple relationship between numbers, there is little room for complications. For example, let's look at a savings account. There are exactly three protagonists: *capital*, *interest rate*, and *interest*.

In any problem about a simple savings account, you are given two of these protagonists and you have to find the third one. That gives us the three essentially different percent problems. They are:

Case 1.
Capital $3,000   Interest rate 6%   *Interest?*

Case 2.
Capital?   Interest rate 6%   Interest $180

Case 3.
Capital $3,000   *Interest rate?*   Interest $180

Now let's solve the only three essentially different percent problems in the whole wide world:

CASE 1.

Capital $3,000   Interest rate 6%   *Interest?*

One percent of $3,000 is the same as $30. (Kill two zeros.)
Six percent of $3,000 is the same as $180.
Done!

CASE 1.3

Capital $3,000   *Interest rate?*   Interest $180

One percent of $3,000 is $30. What percent of $3,000 is $180? (You are asking yourself: How many percents [$30] are in $180?)
The answer is 6%.

CASE 3. 2

Capital?   Interest rate 6%   Interest $180

Here things look bleak at first. We don't know the capital. But wait! We know that:

> 6% of capital is $180
> 1% of capital is $30
> 100% of capital is $3,000 (add two zeros to $30)

You have learned the most important part of practical math. Percents dominate the everyday math you need in stores, in managing your personal finances, in reading and giving information on your job. Budgeting, planning, determining the soundness of your investments, and monitoring the health of the economy are all measured in percents.

SESSION 4.

DECIMALS

Learning math is like learning to write or run or play tennis. Fifteen minutes a day of practice won't do it. You need a new-comer's enthusiasm.

When I started running, my whole life changed. My self-image was first; seeing myself as a serious athlete was a total breach with the past. To my amazement, I started reading about nutrition, I did stretching exercises, I noticed people's stride and gait in the street, I learned about the best jogging paths in New York City. I had always walked with my head angled down at 45 degrees. Now that I was running distances, I lifted my head and saw the tremendous space opening up and down Manhattan's avenues, the clear view of the horizon. That's what I want for you. Learning math is not a one-two-three business like learning little tricks. It means reading things you ordinarily would not look at—a science magazine, the business section of the newspaper, even if you don't understand 10 percent of it— risking a conversation with somebody who is into math, figuring puzzles and problems, diving into it. When I started teaching math-anxiety clinics and formulating my ideas and techniques, I needed something in my life to which I could apply those ideas. I noticed that *The New York Times* crossword sent me into the same kind of a tailspin as that experienced by math-anxious people. I would say to myself, "I'm not a native speaker. I missed all the early days of radio and television. I don't know the local lore." But I used some of my confidence-building techniques and soon I was doing the crossword in pen like all the

113

people I had admired on the subway. I did not set out to acquire tremendous amounts of information, like the people who train for "Jeopardy." What changed was my attitude and my method of attack. *I stopped worrying* about exactly how I should proceed with the puzzle. *I guessed* at some of the answers. *I started* at the beginning or in the middle—*anywhere. I put down* the puzzle when I got stuck in the morning *and picked it up later* in the evening. Amazing how many clues fell into place at 5:00 P.M. that had elicited no response at 9:00 A.M. Be bold, and go for it! All the attack methods mentioned above are very useful when doing math.

You already know what's coming: chain problems! (Yes, they'll have a few percents.)

MENTAL ARITHMETIC

1. 16×25	2. 10% of 1 million	3. 30×30
take 10%	take 1%	take 2%
$\times$ itself	$\div$ 2	take 50%
take 10%	take 5%	$\times$ 50
take 20%	$\times$ 8	take 20%

The answers are (1) 32; (2) 200; (3) 90.

The third chain problem worked out: $30 \times 30 = 900$; 2% of $900 = 18$; 50% of 18 is 9; $9 \times 50 = 450$; 20% of $450 = 90$ (It's easier to do 10% first: 10% of $450 = 45$!)

If you had trouble with the percents, just remember that 1% means simply 1 on every 100, or 1 cent on the dollar.

DARE TO DO DECIMALS

Decimals are not new to you. You have been dealing with money for a long time, and our money is written down in decimal notation (the Old English pound-and-shilling currency was not decimal).

When we refer to fifteen dollars and seventy-nine cents, we write $15.79. The way this is written is called "decimal." The decimal point marks the boundary between whole dollars and cents. The position values are as follows:

Notice the decimal point does not occupy a column.

Decimals beyond the two positions that we are used to from dealing with money occur in science, where precision is extremely important, and in banking. Have you noticed that some banks post interest rates to several decimals? You may see 7.1793% posted as an "effective yield." You know from your percent session that that means 7.1793 cents on the dollar. For small money amounts, these many decimal positions make little sense; for millions of dollars, however, they matter.

HOW TO WORK WITH DECIMALS

ADDITION

Line up the decimal points and fill in the zeros if it makes you more comfortable.

EXAMPLE:

1.72 + 2.8 + 3.705 is rewritten vertically as:

$$
\begin{array}{r}
1.72 \\
2.8 \\
+3.705 \\
\hline
8.225
\end{array}
\quad \text{or} \quad
\begin{array}{r}
1.720 \\
2.800 \\
+3.705 \\
\hline
8.225
\end{array}
$$

SUBTRACTION

This works the same way as addition. Filling in the zeros is helpful here.

EXAMPLE:

7 − 4.745 is rewritten as:

$$
\begin{array}{r}
7 \\
-4.745 \\
\hline
2.255
\end{array}
\quad \text{or} \quad
\begin{array}{r}
7.000 \\
-4.745 \\
\hline
2.255
\end{array}
$$

MULTIPLICATION

You multiply decimals as you would whole numbers. After that, the only question is:

Where do I put the decimal point?

Let's try.

$$
\begin{array}{r}
4.5 \\
\times 4.5 \\
\hline
225 \\
180 \\
\hline
2025
\end{array}
$$

Now where does it go?

First method for placing decimal point: *Look! Estimate!*

4.5 × 4.5 is just about the same as 4 × 5 = 20; so the answer must be 20.25.

This method is reliable and safe, and keeps you in charge and thinking on your feet throughout the problem.

The second method is the way you learned in school: Add up numbers of positions after the decimal points, and cut that many off the answer:

$$
\begin{array}{r}
4.\,⑤ \\
\times\ 4.\,⑤ \\
\hline
2\ 2\ 5 \\
1\ 8\ 0 \\
\hline
20.\,②⑤
\end{array}
$$

In this case, both numbers to be multiplied had *one* decimal position. Add them up and you get two. You cut *two* decimal positions off the answer. This always works, but people tend to think less with this method; they go on automatic pilot and tend to make more mistakes than with the first method.

DIVISION

EXAMPLE:

$$.25\overline{)4.75}$$

If somebody asked you that, do you see that one easy way to do it would be to ask: How many times does 25 cents fit into 475 cents?

What we have done is to convert everything into cents, which we can do easily. Now look at it:

$$\begin{array}{r} 19 \\ 25\overline{)475} \\ -\underline{25} \\ 225 \\ -\underline{225} \\ 0 \end{array}$$

If a division looks more complicated than the one above, pull out your calculator. Your little machine will do all this for you. It is totally dedicated to decimals. The only thing to remember is the keying sequence. For the above example, it would be:

4.75 $\boxed{\div}$ .25 $\boxed{=}$

The answer will be 19.0000.

EXERCISE:

.0063$\overline{)16.45}$ (Key in 16.45 $\boxed{\div}$ .0063 $\boxed{=}$.)

The answer will be 2611.11.

NUMBER SENSE

You have number sense when you have a feel for the answer without working things out. Number sense is a little like a green thumb, and, like a green thumb, it can be trained and developed. It takes *chutzpah* and a sense of thinking on your feet to hit at the answer from afar and trust yourself to get it. Often you may not be able to explain how you got the answer. You just know.

The talents shown by the Dustin Hoffman character in the movie *Rain Man*, and by some so-called idiot savants, are not necessary (and not really desirable). What I would like you to do, however, is use your common sense, life experience, and confident estimating to *avoid* long calculations and serious mistakes.

THE MAGNITUDE PROBLEM, OR HOW MANY ZEROS?

People make the most frequent mistakes with their zeros. Once the numbers become big, people forget to count them: 30 times 30 becomes 9,000 without many even turning their heads, and those mistakes really matter. Even with calculators, people very often make a mistake of one position or two. Teachers sometimes say, "You know how it works, you just misplaced the decimal point," and that's supposed to be reassuring. It is, but in a way those mistakes are terrible, because in real life a factor of ten can kill you. If the nurse makes a mistake by a factor of ten in medication, it could be your end. Or a bridge could fall down, as actually happened in Switzerland a few years ago. So we must learn to pay attention to *magnitude*. What we need is what I recently heard referred to as "a robust sense of numbers."

ESTIMATING

Estimating is a question of number sense, practice, remembering to do it. It ensures accuracy in an important way.

If you have allowed your number sense to bloom, you estimate automatically, every time you deal with numbers. You would never let a misplaced decimal point go because it would just plain *feel* wrong. Consider the following problem:

$$627.97 \times 2.9 = 18,211.13$$

Estimating ($600 \times 3 = 1,800$) tells us that the answer 18,211.13 is ten times too big. Looking at the estimate, we know it has to be 1,821.113.

If this is still foreign to you, practice it all day long. It will do wonders for your sense of competence. Many people don't trust estimating. Let's do a few examples *both ways*:

EXAMPLES:

1. 12.05 × 1,997.27
Estimate: 12 × 2,000 = 24,000
Exact answer: 24,067.1035

2. $6,875.23 ÷ 21.17
Estimate: $7,000 ÷ 20 = $350
Exact answer: $324.76

Commonsense rounding off and estimating is a reliable way to assess the magnitude of your results.

A private client of mine who worked for one of the major brokerage houses kept putting the decimal point in the wrong place and lost the company $50,000 in one week. So we can't be casual if there is an extra zero. If you don't get the exact numbers up front, if it's 916,000 instead of 915,000, that's not going to ruin your company, but if you quote a price or estimate of $91,500 instead of $915,000, where the decimal point goes is absolutely crucial.

People don't have a clear feeling, particularly of big numbers. Let's look at one million. If I say "one million," what happens in your mind? Do you see a 1 with six zeros? That's one good way to go. Do you see it differently? *Million* is such a packaged word, representing such a monolithic number, that a lot of people don't deal well with it. Is there any other way to see it?

I find it is much better to see a million as 1,000 × 1,000. Imagine a pile of a thousand $1,000 bills. Then it's easy to manipulate it. Half a million is 500,000; a quarter of a million is 250,000.

People have a lot of questions about zeros, such as "When you're doing fast math in your head, is it a rule of thumb that if 30 × 30 has two zeros, does 30 × 300 automatically come out to a number with three zeros?"

The answer is no. You have to think things through. Let's

look at this, because sometimes you get extra zeros. Consider 20 × 50, just to show you what I mean. You cannot say the final number is going to have two zeros, because what is 20 × 50? It is 1,000. You see there will be three zeros.

MULTIPLYING NUMBERS THAT END IN ZEROS

Step 1. Unhook zeros
Step 2. Multiply
Step 3. Put all zeros back, attaching them to your answer

EXAMPLE 1: 30 × 400 = ?

Step 1. 3 × 4 (unhook 3 zeros)
Step 2. 3 × 4 = 12
Step 3. 12,000

EXAMPLE 2: 20 × 50 = ?

Step 1. 2 × 5 (unhook 2 zeros)
Step 2. 2 × 5 = 10
Step 3. 1,000

This method keeps you safe with very large numbers, where our number sense is no longer very accurate.

SESSION 5.
FRACTIONS

Many people think of math as a fixed body of knowledge. I don't know whether they think it fell from the sky whole or was given to Moses on the stone tablets, but very few people understand that math developed slowly from primitive roots. In some primitive societies, any number above three is represented only by the word "many," roughly corresponding to the perception of a two-year-old child, in our culture.

Math has evolved along with humanity. The way math is done depends on culture. Just look at a book on the history of mathematics, and you will see. People often have no idea that, among numbers, fractions are extremely old; decimals and percents are much younger. The science of geometry is incredibly old, but algebra is much younger. Seeing math as a human endeavor such as music or painting helps to make it seem less intimidating and rigid. Math changes and expands all the time, and whole new fields have been developed in this century alone.

Yes. You know what's coming. Chain problems!

MENTAL ARITHMETIC

1. 5 × 5	2. .5 × .5	3. .1 × .1
× .1	× 8	× 600
+ 1.1	÷ 4	÷ 4
÷ 6	× 2	÷ .5
÷ 2	÷ 4	× itself

The answers are (1) .3; (2) .25; (3) 9. The third chain problem worked out: $.1 \times .1 = .01 \times 600 = 6 \div 4 = 1.5 \div .5 = 3 \times$ itself $= 9$

FACING FRACTIONS

Fractions are a natural way of expressing numbers that are not whole. The easiest way to get a handle on them is to look at this sequence:

$$16, 8, 4, 2, 1 \ldots$$

To continue, you have to cut 1 in half, then in half again, etc. Imagine a circle:

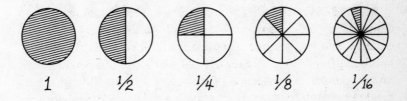

The sequence can now happily continue, thanks to our newfound numbers:

$$16, 8, 4, 2, 1, \tfrac{1}{2}, \tfrac{1}{4}, \tfrac{1}{8}, \tfrac{1}{16} \ldots$$

Note that the bottom number (called the denominator) is doubled every time, resulting in a smaller slice.

The fraction $\frac{1}{8}$ means the whole (thing, capital, price, peach, whatever) has been cut into eight equal pieces of which we take one (one out of eight); $\frac{3}{8}$ means the whole thing has been cut into eight equal pieces of which we take three (three out of eight).

Can you tell what $\frac{1}{6}$ means? Yes, one of six equal parts of a whole. What happens in your mind when you see $\frac{3}{4}$ (pronounced "three-quarters" or "three-fourths")? Most people just see the fraction written, yet it is much more useful to make your mind paint a picture (a right-brain method!). The latter is much easier to manipulate. Let's say you need to double ³/₄:

3/4

³/₄ = ¹/₄ + ¹/₄ + ¹/₄ (three out of four equal parts)

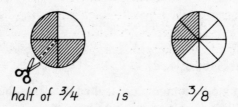

3/4   +   3/4   =   1½

Take half of ³/₄:

half of ³/₄   is   ³/₈

As the picture above shows, half of ³/₄ is ¹/₄ + ¹/₈, which is the same as ³/₈ (three pizza slices),

We have looked at the most important fractions: halves, quarters, eighths. From the pizzeria to the supermarket to the hardware store, these fractions will carry you for a while. Fractions also appear when you are measuring with a ruler or reading gauges. Can you read the positions of these arrows?

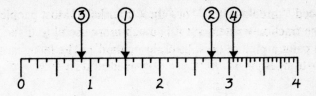

Answers: ↓1 = 1½; ↓2 = 2¾; ↓3 = ⅞; ↓4 = 3¹/₁₆

ADDITION/SUBTRACTION

SIMPLE FRACTIONS

I'll stress the very simple fractions because they're the ones that come up in the stock market, the cookbook, or the hardware store. People's fear of fractions makes them expect fractions around every corner. In real life there are few fractions beyond halves, quarters, and eighths.

Adding and subtracting simple fractions can be achieved easily by drawing (and later by imagining these drawings). Try this with the following examples, either by drawing or by imagining that you work in a bakery or pizzeria where pies are cut into quarters and eighths:

EXAMPLES:

1. ½ + ¼ = _____   6. 1½ − ¾ = _____
2. ½ + ⅛ = _____   7. 1 − ⅜ = _____
3. ⅜ − ⅛ = _____   8. 2¼ + 4⅛ = _____
4. ½ − ⅜ = _____   9. 7¾ − ⅝ = _____
5. 1½ + ¾ = _____   10. 3½ + 6⅞ = _____

Answers: (1) ¾; (2) ⅝; (3) ¼; (4) ⅛; (5) 2¼; (6) ¾; (7) ⅝; (8) 6⅜; (9) 7⅛; (10) 10⅜.

125

Imagine a clock for the following examples:

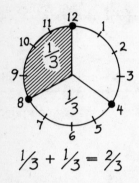

$$\frac{1}{3} + \frac{1}{3} = \frac{2}{3}$$

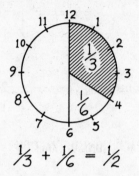

$$\frac{1}{3} + \frac{1}{6} = \frac{1}{2}$$

Many people like the drawings but feel self-conscious about using them at work. Be pragmatic. If it works, use it. Your colleagues will either pay no attention or might be intrigued. After a while, with practice, the drawings will appear in your mind and need no longer be committed to paper.

MORE INVOLVED FRACTIONS

If your work demands ease with fractions such as $7/17$ and $13/69$, you need to generalize the idea of pies and stretch your imagination to include 69 little slices of a pie. Or go on and read a detailed arithmetic book. Math exams, too, still ask for this kind of arithmetic.

Hint 1: Often it is profitable to convert "ugly" fraction problems into decimals and work them on the calculator. (See page 146, on converting fractions to decimals.)

Hint 2: There are scientific calculators on the market that actually do fractions.

I want to emphasize, however, that most areas of life are amazingly fraction-free!

MULTIPLICATION

Fractions were made for multiplying, and that's what comes up most in daily life.

"A quarter-pound of butter . . ."

"He gave three-quarters of his money to charity . . ."

"The dog population of New York City equals one-seventh of the human population . . ."

"A fraction of" mathematically means a fraction multiplied by another number:

$$\frac{1}{2} \text{ of } 8 = \frac{1}{2} \times 8 = 4$$
$$\frac{1}{4} \text{ of } 20 = \frac{1}{4} \times 20 = 5$$

Natural numbers and our early experiences with them hold a lot of power in our mind. The reason people have trouble with this skill is the "faulty" impression the word *multiplication* gives in this case. Do you agree with the following two sentences?

Multiplication makes things bigger.

Division makes things smaller.

In the Book of Genesis, Adam and Eve are told to "be fruitful and multiply." Or we say, "They multiply like rabbits." Obviously these words and images stay with us. Common English usage supports our early ideas of multiplication. But multiplication acts differently with fractions. When you multiply by a fraction of a number, the answer will come out *smaller* than that number.

Multiplication of fractions works this way:

$$\frac{1}{4} \times \frac{3}{4} = \frac{3}{16}$$

Multiply the top numbers: $\quad 1 \times 3 = \dfrac{3}{16}$
Multiply the bottom numbers: $\quad 4 \times 4 = $

Note: Every whole number can be written as a fraction with a denominator (bottom number) of 1. For instance, 4 can be written as $\frac{4}{1}$; 10 can be written as $\frac{10}{1}$.

EXAMPLES:

1. $\frac{1}{2} \times \frac{1}{2} = $ _____
2. $\frac{1}{3} \times \frac{1}{2} = $ _____
3. $\frac{3}{4} \times \frac{3}{4} = $ _____
4. $\frac{1}{4} \times \frac{1}{2} = $ _____
5. $\frac{1}{2} \times 2\frac{1}{2} = \frac{1}{2} \times 2 + \frac{1}{2} \times \frac{1}{2} = $ _____
6. $\frac{1}{4} \times 16\frac{3}{4} = \frac{1}{4} \times 16 + \frac{1}{4} \times \frac{3}{4} = $_____

Answers: (1) $\frac{1}{4}$; (2) $\frac{1}{6}$; (3) $\frac{9}{16}$; (4) $\frac{1}{8}$; (5) $1\frac{1}{4}$; (6) $4\frac{3}{16}$.

DIVIDING FRACTIONS:
WHO GETS FLIPPED AND MULTIPLIED?

Do you have an image of what $8 \div \frac{1}{4}$ means? Most people don't. Let's detour for a moment:

Eight divided by four is clear. Division by four means "cut into four equal parts." But that definition is meaningless for fractions. So let's look at the alternative definition for $8 \div 4$: "How many fours are in 8?" That definition can be expanded for fractions. So, $8 \div \frac{1}{4}$ means, "How many $\frac{1}{4}$ (quarters) in 8?"

Fred says "Thirty-two," without hesitation. How did he do that? "Oh, it's easy," he said. "There are four quarters in a whole. Four times eight is thirty-two."

Notice what he did: $\frac{1}{4}$ became 4. Division became multiplication.

That is exactly the rule that goes through people's minds: When dividing by a fraction, invert it and multiply.

CAMEL PUZZLE

Once upon a time there was a sheik who had three sons. When he died, his sons were informed of his last wish. The sheik's seventeen camels were to be divided among the three sons in the following way: the eldest was entitled to one-half the camels, the second son to one-third, and the youngest to one-ninth. The sons were understandably puzzled, and decided to ask their father's wise old friend for help. The wise man smiled, picked one of his own camels, and added it to the seventeen. Now there were eighteen camels. The eldest took half, or nine camels; the second son took one-third, or six camels; and the youngest took one-ninth, or two camels. Seventeen camels in all. What was left over? The wise man's camel! The old man smiled, wished the sons good luck, and went home with his camel. Can you explain how this happy ending came about?

Let me give you a hint. The trouble is with the particular fractions. To see the situation clearly, I'll give you a simpler version:

The old man had seven camels. The first son gets half of the estate, the second gets a quarter, the third gets an eighth. Now the same problem crops up. How can you take a half of seven, or a quarter or an eighth! We don't want to butcher the poor beasts. But an extra camel rides into town—hooray, hooray!—and that makes eight. Now we can divide the estate. The first son gets four camels, the second gets two, the third gets one. That means a total of seven camels are distributed, and the extra camel rides out of town, a little miffed but otherwise okay.

So we executed the will just fine, but why did we need the extra camel? Let's draw. Given the fractions in the new version of the puzzle, draw a pie and shade in the fractions for the first son, the second son, and the third son.

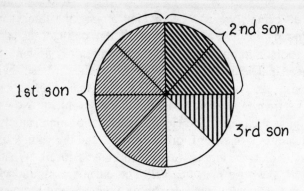

Now you will see that there is a problem.

The shares, the fractions, do not add up to the whole. A half and a quarter and an eighth are not a whole, and so it's an insane way of dividing an estate. That's why, by bringing in the eighth camel, you can divide the shares and have one left over, and that camel can go back home.

In the early version with seventeen camels, the fractions are $\frac{1}{2}$, $\frac{1}{3}$, and $\frac{1}{9}$, and you will find that they add up this way: $\frac{9}{18} + \frac{6}{18} + \frac{2}{18} = \frac{17}{18}$.

Now you see why the wise man's camel had to serve as a stand-in to facilitate division. Nobody wanted to take fractions of camels!

■　　■　　■

SESSION 6.
FUN ALGEBRA AND
BRICK ARITHMETIC

Many clients have told me that their minds literally hide from them when they attempt to do math. Yet the same brain is capable of extremely complex computations that lie outside of our awareness. Your brain is no exception. Every time you play tennis or cross the street in heavy traffic, your brain handles speed, time, and distance, and coordinates them with an elegance and accuracy that would make you pale. It is a fact that when you cross the street without benefit of a traffic light, you must estimate the speed and distance of the oncoming cars in both directions, compute your own walking speed, coordinate all three, and then decide whether you're going to cross or not, and all the time you're thinking about making dinner. I think that's pretty amazing. To figure these things on paper, you would need a certain amount of physics and possibly some calculus. Our subconscious is extremely sophisticated. If you're able to make your way to work every day, you're certainly capable of handling any math that is expected of you.

What's coming? Of course: chain problems!

MENTAL ARITHMETIC

1. $5 \times 5 \times 5$	2. 3×200	3. 4×50
$\times 4$	take 30%	take 10%
take 8%	$\div 4$	$\times 8$
$\times$ itself	$\div 5$	$\div 4$
take 1%	$\times$ itself	take 25%

The answers are (1) 16; (2) 81; (3) 10.

FUN ALGEBRA

Look at the following examples. See if you can fill in the missing number to make things work out. Most people enjoy doing this. It's like a mystery. Was it Jean-Paul Sartre who said people have a deep need to fill in every empty space? Go!

EXAMPLES:

1. $6 + \square = 12$
2. $\square + 3 = 21$
3. $3 \times \square = 24$
4. $\square - 2 = 8$
5. $2 \times \square + 5 = 17$
6. $2 \times \square = \square + 8$
7. $\square \times \square = 25$
8. $\square \times \square = 36$

Answers: (1) 6; (2) 18; (3) 8; (4) 10; (5) 6; (6) 8; (7) 5; (8) 6.

The concept of the empty box works nicely for equations, where we just slot in the number that makes things work out. Let's look at a little word problem:

Jim is twice as old as Toby. Together they are 36 years old. How old is Toby? I like those boxes from before. They seem to serve well to write down a number we don't know. Since we have to build a picture of the situation here, I prefer the word *brick* for the empty box. I'll call them bricks from now on—it is such a solid, constructive concept.

Math consists of a lot of "as ifs." We pretend we know

Toby's age (we call it a brick) and then we write down the information:

If Toby's age is

Jim's age is (twice as old)

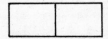

Let's play "choo-choo," combining the bricks for both their ages:

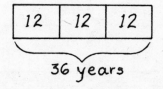

If all three bricks equal 36 years, one brick equals 12 years (36 ÷ 3 = 12), and two equal 24. Toby is 12 years old, and Jim is 24.

Choosing a brick to represent Toby's age enabled us to put in clear context all the information we had, and this helped us solve the problem.

BRICK ARITHMETIC

How did you like those bricks? I like them because they give you something to do when faced with a problem. You have no time to stare into space or get nervous. You have to draw bricks.

Let's try another example:

John brags about his 25% raise-cum-promotion-pay-increase. Later that night he says, "Boy, I'll make $30,000 more next year." How much is he making before the raise? After the raise?

Who gives the pay raise? The boss. Let's get into the situation: Let's be the boss. Take John's salary. We don't know how much it is, so we act "as if" we did, and represent it by a bar:

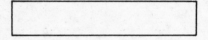

The raise is 25% or one-fourth of the salary.
The old salary is four bricks, each 25%.
The new salary is five bricks.

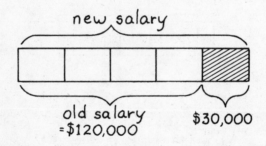

That wasn't so bad, was it?

Vanessa came to see me one day with this question:

If Joe's tax bracket is 25% and his *net* income is $24,000, how much is Joe's salary?

Vanessa asked, "Isn't there something with *x*? This is algebra, I can't do algebra!" Vanessa was partly right. These problems are usually taught as part of a conventional algebra course, full of equations and *x*'s. But we will use *brick arithmetic*.

We draw Joe's salary as a big bar:

What happens? Taxes eat up 25% (¼)!

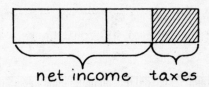

net income   taxes

The net income (three bricks) amounts to $24,000. That means every brick is worth $8,000.

$8,000	$8,000	$8,000	$8,000

So Joe's salary is 4 × $8,000 = *$32,000.*

Vanessa relaxed. She liked the pictures. They made math look more down-to-earth.

At this point I can just hear the skeptics among you leveling questions at me such as, "What if the numbers are more complicated?" "What if the tax rate was 35 percent?" "What if the tax rate was 17 percent?"

You have a point. The good news is that our method works the same way with these numbers. You just need a little more imagination.

Let's do Joe's problem with 17%. (You need a calculator for this.) Joe's salary is again symbolized by a long bar:

What happens? Taxes eat up 17%.

Imagine the bar cut into 100 equal bricks, each representing 1%.

83 bricks for net income   17 bricks for taxes

That leaves 83 bricks for the net salary. The net salary is 83 bricks = $24,000.

One brick . . . wait, let's do this with a calculator. Key in 24,000 ÷ 83 =, and you will see that 1 brick = $289.16, so 100 bricks = $28,916.

Does it make sense to you that Joe's salary this time is smaller than before? Vanessa could not see it at first. She said, "If the tax rate is only 17%, he should have more money left over."

Were you thinking like that?

An explanation in words would take at least two paragraphs. That's where pictures come in handy. Compare:

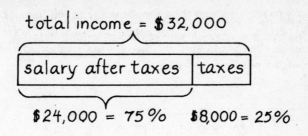

total income = $32,000

salary after taxes	taxes

$24,000 = 75% $8,000 = 25%

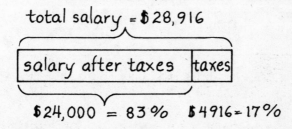

total salary = $28,916

salary after taxes	taxes

$24,000 = 83% $4916 = 17%

Voilà! The tax in the second case is smaller, i.e., the total salary is smaller.

Vanessa's next question concerned percentages over 100:

A store on Fifty-seventh Street in New York City suffered a 300% rent increase over the last five years. The rent is now $20,000 a month. How much was it five years ago?

Vanessa didn't believe there was anything beyond 100%. Wasn't 100% the whole thing, the top, the limit?

It turns out that if you rent on Fifty-Seventh Street, your 1984 rent was topped fast and furiously. Let's draw the basic brick for the old rent (100%):

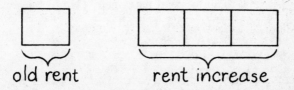

old rent   rent increase

Can you see now that the expression "The rent has increased by 300 percent" means the same thing as that it has increased by three times the old amount or "the rent has multiplied by four." So, if we put them together:

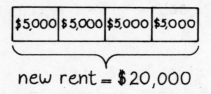

new rent = $20,000

We see that every brick is $5,000. So the old rent is $5,000.

EXAMPLE:

You buy a coat during a 10%-off sale and you pay $180. How much was the coat originally?

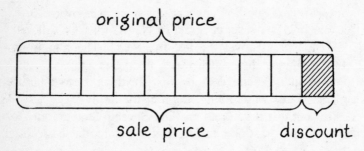

Represent the original price of the coat by a bar. Chop it into ten equal bricks to show what happened. *One* of the ten bricks represents the discount (10% = $^1/_{10}$!).
Nine bricks represent the sale price.
Nine bricks amount to $180.
One brick is then $20.

The discount was $20. The original price is all ten bricks, and that amounts to $200.

I hope you can see that drawing bricks for numbers and for what happens to them clarifies the basic relationships between numbers in a much more obvious way than equations do.

SESSION 7.
PROPORTIONS

Letting yourself make mistakes is essential. Looking at mistakes calmly and backtracking your thoughts give you a map of your mind at the moment. If you can do this, you are not running away from anything and you have a sense of security. If you're not afraid of your inner critic, you will not worry about the critics in real life. Mistakes are shortcuts to learning. In my classroom, I love juicy mistakes. They help to demonstrate clearly what needs to be learned and changed.

I often find that students write small and tentatively in math. "Why do you use a pencil?" I ask.

"Because there is so much erasing," they answer.

"I don't want you to erase even one line of your work," I say. "Show yourself basic respect. You wrote it for a reason. It can stand. It is worth looking at. It helps you to have a sense of process and progress. Every step you take helps you feel grounded."

Here we go again: chain problems!

MENTAL ARITHMETIC

1. 8 × 25
   take 40%
   × 10
   take 1%
   × 11

2. 5,000 ÷ 10
   take 1%
   × 12
   ÷ 4
   × 5

3. 1 million ÷ 1,000
   ÷ 4
   take 10%
   × 4
   × itself

The answers are (1) 88; (2) 75; (3) 10,000.

PROPORTIONS

There is a kind of math question that gives disproportionate trouble to many adults. The questions run something like the following:

1. If 5 pounds of apples cost $3, how much are 7 pounds?

2. David lives on his investments. He says, "The average yield of my investments is 9 percent, and I need about $45,000 a year. How much capital do I have to have?"

3. Jim leaves his job on September 1. How many of his official 18 annual vacation days is he entitled to?

Most people learned to solve this kind of problem as part of an early algebra course. Let's look at number 3. Do you remember this way of expressing such a problem?

Let x = number of vacation days due to Jim; $x/18 = {}^8/_{12}$ (he worked 8 months out of the total 12); cross multiply $12x = 144$; $x = 12$.

Take heart. If this sets your mind spinning, there is an easier way. Let's call it the European way, because it's taught more often across the ocean. You write simple English sentences. You *don't* use *x*. You *don't* write a proportion. *You go back to the basic unit.*

EXAMPLES:

1. 5 pounds of apples cost $3.00
   1 pound of apples costs $3.00 ÷ 5 = 60 cents
   7 pounds of apples cost 7 × 60 cents = $4.20

   (The basic unit can be 1 pound, 1 day, 1 minute, 1%, etc.)

2. 9% of David's capital is $45,000
   1% of David's capital is $45,000 ÷ 9 = $5,000
   100% of David's capital is 100 × $5,000 = $500,000

3. For 12 months, Jim gets 18 vacation days
   For 1 month Jim gets 18 ÷ 12 = 1.5 vacation days
   For 8 months Jim gets 8 × 1.5 = 12 vacation days

PROPORTIONAL LEAPS

Vanessa had trouble doing proportional leaps. I'll show you what that means:

   If 1 pound of apples costs 39 cents, how much are 4 pounds? Do you see that you have to multiply by 4? That's good.

$$\begin{array}{r} 39 \text{ cents} \\ \times\, 4 \\ \hline 156 \text{ cents} = \$1.56 \end{array}$$

Vanessa saw 4 pounds as 3 pounds more, and had trouble seeing the multiplication. Let's do it in reverse:

   If you know that 6 ounces of steak cost $1.80, how much is 1 ounce?

   Do you see that you have to divide by 6? So 1 ounce of steak costs $1.80 ÷ 6 = $.30.

Vanessa said, "One ounce is 5 ounces less." Her statement is true, but it does not help you to get there. Vanessa had trouble "sniffing out" division problems. You knew that you had to divide by 6.

EXAMPLE:

A 6-ounce can of salmon costs $3.60. How much per pound (1 pound = 16 ounces)?

Answer: $9.60.

■ ■ ■

SESSION 8.
CONVERSIONS

When you learned math in school, chances are that you learned it for good grades (or for bad!), just for the next test. Now is your chance to learn math for yourself, for your own sake, for your life, and for use in practical situations. You must also realize that math, like every other human endeavor, needs time, caring, motivation, and lots of practice if you want to become good at it. This is the beginning of your trusting your own mind. A great moment.

I don't have to tell you what's coming: chain problems!

MENTAL ARITHMETIC

1.	2.	3.
20 × 100	50% of 21	10% of 25
take 20%	÷ 2	+ 3½
÷ 10	− ¼	× 15
× itself	× 1,000	take 20%
take 1%	take 5%	÷ 4

The answers are (1) 16; (2) 250; (3) 4½.

Often you may feel confused because you need to understand numbers as fractions one minute and as percentages or decimals the next. I want to stress again that fractions, percentages, and decimals are different *languages* to express the same numbers, just as the word *table* in English is *Tisch* in German and *mesa* in Spanish.

Let's look at the following three pictures:

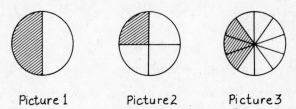

Picture 1   Picture 2   Picture 3

Picture 1 can most easily be seen as ½.
Since the whole thing is always 100%, half of it is 50%.
Since the whole is 1.00 in decimals (think one dollar), half is .50.
So the best thing to remember and memorize is:

$$\tfrac{1}{2} = 50\% = .50$$

Tape it to your desk or your refrigerator, or visualize it in your mind every time you need to translate. Now do the same thing with the other two pictures.

$$\text{(Picture 2)} \quad \tfrac{1}{4} = 25\% = .25$$
$$\text{(Picture 3)} \quad \tfrac{3}{10} = 30\% = .30$$

Notice how decimals and percents are similar in appearance, using the same digits. They differ only by two decimal positions; 50% is .50. That makes sense when you realize that the "%" sign means "divide by 100" ($50\% = 50 \div 100 = .50$).

HOW TO CONVERT

Each of the three languages can be translated into any of the others, so there are six possible ways to translate:

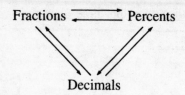

Some paths are more practical than others.

CONVERTING FRACTIONS TO DECIMALS

The fraction bar is an old division sign. Your parents or grand-parents wrote fractions as 1/2 and 3/4. The slash was well known as a division sign, so 3/4 means $3 \div 4$ or $4\overline{)3}$.

The first version is what you need when you key into a cal-culator, and the second when you do long division by hand. In short, the fraction-to-decimal conversion is achieved by dividing!

Let's practice:

1. $4/10$ = ____
2. $7/10$ = ____
3. $1/8$ = ____
4. $5/8$ = ____
5. $1/3$ = ____
6. $5/6$ = ____

Answers: (1) .4; (2) .7; (3) .125; (4) .625; (5) .333 . . . ; (6) .8333 . . .

CONVERTING FRACTIONS TO PERCENTS

A. Easy Conversions: $1/2$, $1/4$, $3/4$, $1/10$, etc.

These easy ones you'll have used so much by the end of this book that you'll see them very clearly without effort. Let's try:

1. $\frac{1}{2}$ = ____
2. $\frac{1}{4}$ = ____
3. $\frac{3}{4}$ = ____
4. $\frac{1}{10}$ = ____
5. $\frac{3}{10}$ = ____
6. $\frac{2}{5}$ = ____

Answers: (1) 50%; (2) 25%; (3) 75%; (4) 10%; (5) 30%; (6) 40%.

B. Less Obvious Conversions

The easiest way to go from fractions to percentages is via decimals. Do the decimal first: $\frac{1}{2} = 1 \div 2 = .50$.

Then move the decimal point two places to the right: .50 = 50% (You remember that. You stuck it to the refrigerator!)

EXAMPLES:

$\frac{1}{8} = 1 \div 8 = .125 = 12.5\%$

$\frac{2}{3} = 2 \div 3 = .6666\ldots = 66.66\ldots\%$

$\frac{9}{10} = 9 \div 10 = 90 = 90\%$

$\frac{1}{20} = 1 \div 20 = .05 = 5\%$

$\frac{7}{9} = 7 \div 9 = .777\ldots = 77.77\ldots\%$

(I have friends who tell me that the .6666 . . . makes them nervous, and they wake up in the middle of the night thinking of all those sixes going on forever.)

CONVERTING DECIMALS TO PERCENTS AND PERCENTS TO DECIMALS

As mentioned above, this is the easiest transition. Remember:

$$50\% = .50$$
$$.50 = 50\%$$

Just watch:

75%	= .75	.45	= 45%
80%	= .80 (or .8)	.08	= 8%
7.5%	= .075	.005	= .5% = ½%

CONVERTING PERCENTS TO FRACTIONS

Let's look at percents carefully: "%" means "divide by 100"; so percents are special fractions. They all have a denominator of 100:

$17\% = {}^{17}/_{100}$
$20\% = {}^{20}/_{100} = {}^1/_5$
$5\% = {}^5/_{100} = {}^1/_{20}$
$25\% = {}^{25}/_{100} = {}^1/_4$

CONVERTING DECIMALS TO FRACTIONS

Remember the very definition of decimals. The columns tell us how big a number is, what the value of each digit is.

$.1 = {}^1/_{10}$	$.7 = {}^7/_{10}$	$.8 = {}^8/_{10}$
$.01 = {}^1/_{100}$	$.07 = {}^7/_{100}$	$.28 = {}^{28}/_{100}$
$.001 = {}^1/_{1,000}$	$.007 = {}^7/_{1,000}$	$.287 = {}^{287}/_{1,000}$

CONVERTING REPEATING DECIMALS TO FRACTIONS

There is no simple way to make infinitely repeating decimals into fractions, but the same ones come up again and again, and we just learn to recognize them:

.3333 ... = $\frac{1}{3}$		.1111 ... = $\frac{1}{9}$	
.6666 ... = $\frac{2}{3}$		.2222 ... = $\frac{2}{9}$	
		.3333 ... = $\frac{3}{9}$ = $\frac{1}{3}$!	
.16666 ... = $\frac{1}{6}$		.4444 ... = $\frac{4}{9}$	
.8333 ... = $\frac{5}{6}$		.5555 ... = $\frac{5}{9}$	
		.6666 ... = $\frac{6}{9}$ = $\frac{2}{3}$!	
		.7777 ... = $\frac{7}{9}$	
		.8888 ... = $\frac{8}{9}$	

I have often had students jump up and down and get all excited at the mention that in $\frac{1}{3}$ = .333 . . . the threes repeat themselves ad infinitum. The very mention of infinity touches off strong feelings, because we cannot see or count or write down all those digits. If you need everything around you under your control, this concept is bound to make you a little uncomfortable.

■ ■   ■

SESSION 9.

ORIGAMI

Math is abstract—maybe. Whenever I start teaching a class, I ask my students about their interests, hobbies, and experiences. This is not politeness. I know that baking, sailing, weaving, farming, and playing racquetball and tennis support a person in learning math. We tend to underestimate how important our bodies are in learning. Little children, for instance, learn a lot more when they are allowed to count on their fingers or count objects rather than moving too quickly to abstract numbers. The same is true for us. If we work things out, if we have a geometry problem and we fold paper and if we draw, if we use our hands and possibly our whole body, I know from my training in body-work that it does change our thinking patterns and our nervous systems.

In my workshops I ask people to measure the span of their hands. I know that the span of my hand is seven inches. I know how long it is from the tip of my middle finger to my elbow. I know how tall I am. I know my full arm span. This is not just to be cute. This is helpful. Say I'm in a furniture store and I need a table to fit in a thirty-inch space. I know that my hand span is seven inches, so I can very quickly measure what I need. All those little concrete skills give us a sense of competence and power. Being concrete and close to things is definitely under-valued in this educational system, and I cannot overstress its importance.

Here we go again: chain problems!

MENTAL ARITHMETIC

1. 10×70
   take 30%
   + 90
   take 25%
   take 20%

2. 8×25
   take off 10%
   ÷ 4
   × 100
   take 20%

3. $1,000 \times 1,000$
   take off 40%
   ÷ 1,000
   take off 10%
   take 10%

The answers are (1) 15; (2) 900; (3) 54.

DOING ORIGAMI

Origami is the Japanese art of paper folding. I'm sure you have seen the elegant birds, frogs, and boats that are made in origami. You start out with a square piece of paper. It is nice to get actual origami paper; it is cut exactly square and comes in beautiful colors and patterns. If you can't find it, just make yourself square paper for the following exercises. Take a regular 8½-by-11-inch piece from a pad, and cut it down to a perfect square. How could you do that?

Do it before reading on.

In my workshops, Janet ponders while Naomi casts a quick glance at her paper, folds up a reasonable-looking flap of paper, and quickly tears it off along the fold.

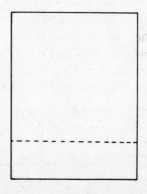

Janet still ponders.

Jim takes out a ruler, measures the top of the paper (8½ inches), marks off 8½ inches along the longer side, and then cuts out his square.

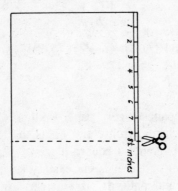

Janet smiles, and turns down one corner of her paper diagonally.

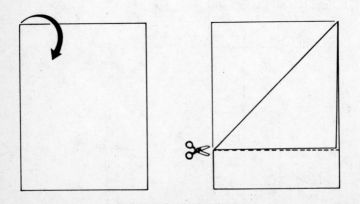

What has she done?

She has used the top edge of her paper to measure off an equal distance on the side. Smart? Yes, no ruler needed!

Any way you did it, you are now in possession of a basically square piece of paper. Can you fold this square paper in a way that will give you a new square, half the size of your original square?

Put the book down, go play, fold, unfold, talk to yourself: What does "half the size" mean? I'll show you what some of my students usually come up with:

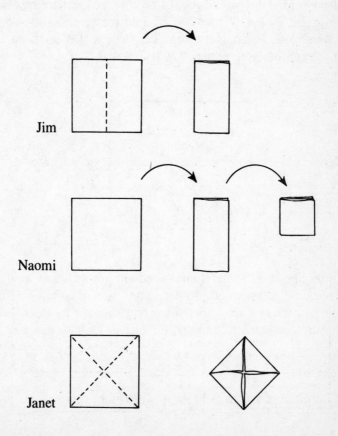

153

Which one did you do? Let's discuss them. They all have good points. Jim's is definitely half the size (size usually refers to area if we're dealing with two-dimensional objects like squares or rectangles), but is it a square? No, it is a rectangle. Jim satisfied only one of the two conditions, which are that it has to be (1) a square and (2) half the size.

Naomi was more impressed with making a square, which her product obviously is, but is it half the size? Look.

Can you see that Naomi's square would fit into the original square exactly four times? So again, it's not the answer.

Janet folded the two diagonals (an easy way to find the center point), and folded all four corners into the center. Did she get a square? Yes. Is it half the size? Let's check. Do what she did, and then unfold your paper. It will look like this.

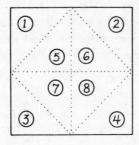

Triangles 1, 2, 3, and 4 combined are just as big as triangles 5, 6, 7, and 8 combined. So her square is exactly half the size. Note how every correct *and* incorrect solution gave us something to think about. The main thing is to let it go and risk the answer.

EXERCISES:

Take a piece of paper and fold it in half four times.

1. How many layers does the wad have?
2. Unfold the wad. How many small rectangles have you created?
3. What fraction is each of these little rectangles' area of the original paper?
4. How many times would you have to fold the paper to create a wad of 64 layers? (Start with a big piece of paper.)

Answers: (1) 16; (2) 16; (3) ¹/₁₆; (4) 6.

AREA AND PERIMETER

Take a piece of origami paper, or make a square paper for yourself. Do what Naomi did earlier. Fold the square in half and in half again, so that you get a smaller square.

Look at the small square. Let's think about area and perimeter. We have already learned that the area of the small square is ¼ the size of the original one. Now what about the perimeter? The word *perimeter* means literally "around-measure." It is the length all around the shape, like a trim or a fence.

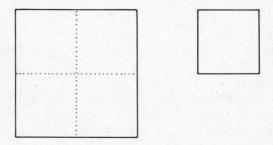

Can you see that the perimeter of the little square is *half* the perimeter of the big one (each new side is half as long; so the whole perimeter is half as long).

If this seems too complicated, measure your square. Let's say your original square is 8 inches by 8 inches. How far is it all around?

Perimeter (large square): $8 + 8 + 8 + 8 = 32$.

Perimeter (little square): $4 + 4 + 4 + 4 = 16$ (half as much).

EXAMPLES:

1. If Ellen's kitchen is 20 by 20 feet and John's is 10 by 10 feet, is Ellen's kitchen twice as big?

2. Farmer Karl has a garden 30 by 50 feet. Farmer Mary Jo has a garden 40 by 40 feet. Compare areas (size) and perimeters (fences).

Answers:

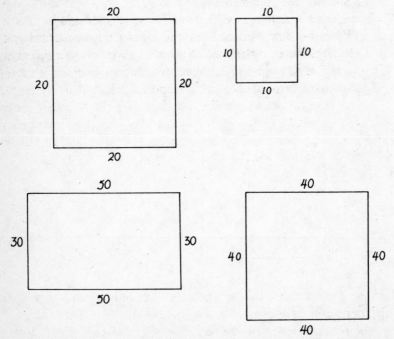

Note: There is no simple connection between perimeter and area. The same perimeter, 160 feet, for instance, can hold large and minuscule areas. The maximum area possible is always a circle. Imagine 160 feet of string laid out in a circle. That's the most area you can capture. There is no lower limit for the area. Take 160 feet of string and lay it out like the outline of a comb, for instance:

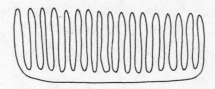

You can see that all the perimeter gets "wasted" on a very small area.

You have now learned two very important geometrical concepts: perimeter, which is the sum of all the sides, and area, which you figure by multiplying length times width. We have talked only about squares and rectangles, but they are by far the most frequent shapes you will have to deal with. Whether you're ordering plastic or wood at work, or whether you're recarpeting your living room and sewing draperies, squares and rectangles will be the dominant shapes. Hooray!

■ ■ ■
WORD PROBLEMS

I want to help people talk their way through problems. Many people find that after reading a problem their minds are shrouded in total silence. They wait for the answer to show up by itself. They misperceive what people who are good in math do to get to the answer fast. Their answer springs from a process practiced hundreds of times:

1. Weighing and using given data
2. Guessing and hypothesizing
3. Checking for plausibility
4. Going back and checking

Since we think at the speed of light, a lot can happen in one second in a practiced mind, and that's what we're going for here, a practiced mind. Math is not located in some secret cerebral lobe that is missing from your brain. It is this belief that makes you think that math always comes easy to those whiz kids. It is a bit like assuming that a magician with talent does not have to practice and that gifted athletes shoot those baskets every time they try. It is amazing that although everyone accepts that athletes practice many hours a day, year in and year out, learning math and developing problem-solving skills aren't generally regarded as requiring the same kind of practice. But those skills reach far beyond math itself into your professional life, into your private life. It's a process that is universally useful. Every day when you read the paper or talk to people, math competence is part of you.

Here they come again, for the last time: chain problems!

MENTAL ARITHMETIC

1.	$\frac{1}{2} + \frac{1}{4}$	2.	$5 - \frac{1}{2}$	3.	$3 \times 1\frac{1}{4}$
	$\times 100$		$\times 4$		$+ \frac{1}{4}$
	$\times 2$		$\times 10$		$\times 75$
	$\div 30$		$\div 4$		$\times \frac{1}{2}$
	$\times \frac{1}{2}$		$\times \frac{1}{2}$		$\div 3$

The answers are (1) $2\frac{1}{2}$; (2) $22\frac{1}{2}$; (3) 50.

Word problems seem to be the most hated problems that people remember from their schooldays, maybe because each one looks so new and different from the last one. There does not seem to be a system, or rules to go by. So many people wish that every word problem came with a little set of instructions attached, such as, "Here you have to multiply, and there you have to divide." In this session we learn to deal with problem-solving from a common-sense point of view, because in real life most math questions/problems do approach us as word problems. Rarely (excluding on employment-agency tests or exams) will someone slam a piece of paper in front of you, saying, "Do this worksheet." Word problems are problems that put math into context.

Obstacles to happy problem-solving include the following:

- *Difficulties with reading.* Incredibly competent people who can easily read a paper, a novel, or a management report are often not trained in reading scientific and mathematical information or instructions. The latter is all the same kind of writing: terse, concentrated, not fluid. Every single word matters.
- *Omission of common sense and imagination.* When faced with

159

math, many people move their thinking to a specific and often narrow region of their mind, excluding all practical knowledge and their life experience.

- *Low frustration tolerance.* Frustration tolerance is to be cultivated and treasured. Some people see it as a sign of maturity. I know extremely mature people who lose all semblance of maturity when faced with a word problem. They read the problem hastily, charge right ahead to answer it, can't find the answer, and are totally confused and frustrated.

Let's look at four open-ended word problems. I call them "stories."

STORIES

1. Sonia's raise and promotion amount to a 25% pay increase.
2. John is twice as old as Gillian. Gillian is twice as old as Toby.
3. A train leaves Paris, traveling westward at 50 miles per hour. A second train leaves, going eastward at 60 miles per hour.
4. In a purse there are dimes and quarters, twice as many quarters as dimes.

I'm sure that some of these stories bring back waves of memories and apprehension. Notice, however, what is special about them. They are open-ended—no questions are asked. Make a list of questions of your own. You can make them easy or difficult. (Be good to yourself; make up easy ones.)

Asking your own questions puts you in charge. You'll see that answering an easy question will help with more difficult ones. I'm listing for you some of the simpler questions my clients came up with. (I'm keeping the tough and exotic ones to myself!)

STORY 1

a. Find Sonia's new salary if her old salary was $20,000.
b. Find her old salary if her new salary is $30,000.
c. Find her new salary if her raise is $8,000.

STORY 2

a. Find all their ages if Toby is 10 years old.
b. Find all their ages if they add up to 35.
c. Find all their ages if the age difference between John and Gillian is 18.

STORY 3

a. If they both leave at noon, how far apart are they at 1:00 P.M.?
b. If both leave at noon, how far apart are they at 3:30 P.M.?
c. If the westward train leaves at 11:00 A.M. and the eastward train at noon, how far apart are they at 1:15 P.M.?

STORY 4

a. If there are 6 dimes, how many quarters are there? How much money?
b. There are 12 coins total. How much money?
c. The value of all the coins is $6. How many dimes?

Do you like these questions? Are they similar to yours? Try to answer yours first with the help of the upcoming six strategies, and then read on to the solutions for those I listed. But you'll have to wait a moment, because we aren't going to solve these problems just yet. Is that hard for you? Do you need to know the answers immediately? Practice increasing your frustration tolerance. Yes, we'll discuss the answers; yes, you'll learn them; but first we want to look at the best strategies to tackle word problems.

SIX STRATEGIES FOR PROBLEM-SOLVING

Having simple, concrete attack methods for math problems does much to relieve math anxiety and math avoidance. You tend to get anxious when you don't know what to do, so these six strategies will give you tools to use every time you get stuck.

1. Breathe.
2. Read questions in small segments, lifting your head and integrating information at each stop.
3. Draw a picture if at all applicable. Use circles and bricks.
4. Rewrite the problem in your own words.
5. Give an off-the-top-of-the-head estimate of the answer.
6. Simplify the problem and do a simplified version first.

This may seem like too much work for one problem. (There never is enough time.) But just pause and think. Dealing with a problem, picking it apart, kicking it around, is the very essence of solving it. The solution is right at the center of the problem, not somewhere off on a tangent. Very often you will need only a few of these strategies.

HOW TO USE THE STRATEGIES

1. Breathe
Don't laugh. I have observed that people just about hold their breath when they feel stumped or overwhelmed by a problem. Often they pull up their shoulders and tighten their necks, cramping the very area that needs to be free and loose for good

thinking. Your brain is a physical organ deserving the best possible treatment. Tension, cramped posture, and shallow breathing lower oxygen supply and blood flow to the brain! Deep breathing improves your thinking *and* your mood. (See page 74.)

2. Read

Read the problem in small segments, lifting your head and integrating information at each step. Let's read Story 2, on page 160, for instance: "John is twice as old as Gillian. Gillian is twice as old as Toby."

That means the age sequence, from oldest to youngest, is (1) John, (2) Gillian, (3) Toby. Write it down for later reference.

Now you are clearer. If it's not that easy, go back and reread. As obvious as it seems, reading is what many people don't do. Skim, glance, scan, yes; read, no. Reading is *thinking*. Few people understand a math problem after one reading. Most mathematicians read the facts and the question at least twice. They'll go forward and then back again, reading in a zigzag fashion. That is a good thing to learn. Then reformulate the facts and the question in your own words. I cannot overstress that. *You have to make the question and the problem your own.* I have found that *misreading* and *misinterpreting* the facts lead to at least half of the errors people make with math problems. Another obstacle to getting the facts straight is anxiety, which makes numbers dance around and words lose their meaning; but if you focus enough and go step by step, you will be able to get the content of the word problem clearly into your mind.

3. Draw

Do this as often as possible. As we learned in Session 6, "Fun Algebra and Brick Arithmetic," producing a drawing is like making a model of the situation. It aids your right-brain thinking.

A mathematician recently said to me, "I always make a list of all the given facts, but every time I try to hold them all in my mind at the same time, I lose one important detail."

Drawing is an antidote to that kind of strain. A picture holds the relationships and the details better than a list, bringing your visual ability and your imagination into play.

4. Rewrite

It helps to deal with the problem in your own language and with imagination. In solving word problems, I like my students to become involved in the stories themselves, to put themselves in the scene, to imagine things in great detail. What do the people in the word problems wear? From where do those trains that go east and west depart? Are we in Paris? Are we in Chicago? I believe strongly that if you make the problem three-dimensional, colorful, interesting, and personal, your whole right brain gets stimulated and involved. "Right brain" is used here as a metaphor for a thinking style involving intuition, memories, life experience, and a sense of fun.

5. Estimate

If you take a rough guess at the answer first, your confidence will soar. If you understand what the word problem is all about, you can risk a rough guess. That mental activity of guessing points your mind in the right direction. If you have no idea how to make a guess, it pays to go back and reread the problem; you may have missed a fact.

6. Simplify

There may be a stumbling block in the word problem: a train has a head start or there is an extra discount you don't know how to deal with. Kick the offensive part of the story out. Do the problem without the irksome detail. Once you get the answer to the simplified problem, it will then usually be clear to you how to do the more complex problem.

STORY 1

Let's use brick arithmetic for Sonia's pay increase. Her old salary can be drawn as a long bar:

The raise is 25%. That means her old salary gets cut into four equal bricks, and one, the raise, is added on. Now we are ready to tackle the questions.

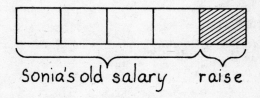

QUESTION 1:

The old salary is $20,000.

$$4 \text{ bricks} = \$20,000$$
$$1 \text{ brick} = \$5,000$$

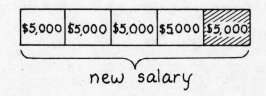

The new salary is 5 bricks.

$$5 \times \$5,000 = \$25,000$$

165

QUESTION 2:

The new salary is $30,000.

$$5 \text{ bricks} = \$30,000$$
$$1 \text{ brick} = \$6,000$$

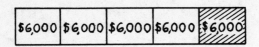

The old salary is 4 bricks.

$$4 \times \$6,000 = \$24,000$$

QUESTION 3:

Raise = 1 brick = $8,000.
New salary is 5 bricks = $40,000.

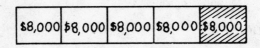

STORY 2

Before we go on to answer any of the questions, let's exploit the information given so far. First, let's use a drawing (brick arithmetic).

Toby is the youngest. Then, from the story:

1 brick = Toby's age

2 bricks = Gillian's age

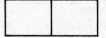

4 bricks = John's age

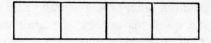

Now we are ready to read the questions.

QUESTION 1:

Find all their ages if Toby is 10. Look at the picture. If Toby is ten, then every brick = 10.

Toby 10, Gillian 20, John 40

QUESTION 2:

Find all their ages if they add up to 35. All the ages in the picture total 7 bricks. 7 bricks = 35 years. That means every brick = 5.

Toby 5, Gillian 10, John 20

QUESTION 3:

Find all their ages if the age difference between John and Gillian is 18 years. *Notice* how complicated this word problem would have sounded to you only a little while ago. Now we have a picture to look at and hold on to.

John's age = 4 bricks, Gillian's age = 2 bricks

Difference in their ages = 2 bricks = 18. That means every brick = 9 years.

Toby 9, Gillian 18, John 36

STORY 3

Let's look at this. Most people say trains are bad enough, but east and west and one train having a head start! It's too much.

Let's take the questions one by one *after* we use the best strategies. *Reading* carefully and *drawing* seem best.

The original situation could be depicted this way:

Now we feel a little better. Let's *rewrite* "50 mph." Most people read *mph* as if it were one word, and a mysterious one at that. Yet all it means is "50 miles every hour." Let's draw an hour-by-hour chart for the trains:

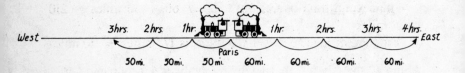

You simply draw out what happens hour by hour, and notice that this drawing can be made independent of the later questions. Now we're ready to answer the questions.

QUESTION 1:

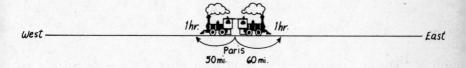

You can "read off" the answer from the drawing. It is

$$60 \text{ miles} + 50 \text{ miles} = 110 \text{ miles}$$

QUESTION 2:

Again, our drawing contains all the information:

The first train covers $50 + 50 + 50 + 25$ miles $= 175$ miles.
The second train covers $60 + 60 + 60 + 30$ miles $= 210$ miles.
The total distance between the trains is 175 miles $+ 210$ miles $= 385$ miles.

QUESTION 3:

The head start need not throw us off. We'll *simplify* the question and start both trains at noon. Draw the situation.

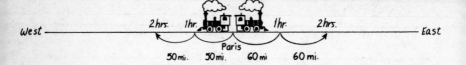

The first train covers 50 + 12½ (or ¼ of 50) = 62½ miles.
The second train covers 60 + 15 (or ¼ of 60) = 75 miles.
The total distance between the trains is 137½ miles.

Done without the head start!

Now what's the trouble with the first train leaving at 11:00 A.M.? It simply means that the first train is running one extra hour, from 11:00 A.M. to noon. During that hour, it covers 50 miles. So the total distance between the trains (with head start) is 50 miles more than before and that is 187½ miles.

STORY 4

This problem calls for very careful *reading*. Many people stumble over "twice as many quarters as dimes." Pause. Does it mean twice as much money or twice as many coins? *Read* again.

Rewrite. The difficult phrase can be replaced by "there are two quarters for every dime." Now this is clearer and we're ready to *draw*.

or a pile:

So the coins can be grouped into little piles worth 60 cents each. Now we're ready for the questions.

QUESTION 1:

If there are 6 dimes, that means there are 6 piles:

$$6 \times 60 \text{ cents} = \$3.60$$

QUESTION 2:

There are 12 coins total. There are 3 coins in every pile. That makes exactly 4 piles. Every pile is worth 60 cents:

$$4 \text{ piles} = 4 \times 60 \text{ cents} = \$2.40$$

QUESTION 3:

The value of all coins is $6. Every pile is worth 60 cents. That means there are 10 piles. If there are 10 piles, then there are 10 dimes.

NOW PRACTICE
PROBLEM-SOLVING ON YOUR OWN

In this section you will find problems that touch on all the topics covered in previous sessions. The emphasis is on learning to approach a problem by *looking* at it. There are methods in problem-solving that can be learned. These methods help you to stay with the problem and the inherent frustrations without tuning out or running away.

REMEMBER THE SIX STRATEGIES

When faced with an unfamiliar problem, there are several things you can do before you have the slightest idea of how to solve it:

1. Breathe.
2. Read questions in small segments, lifting your head and integrating information at each stop.

3. Draw a picture if at all applicable.
4. Rewrite the problem in your own words.
5. Give an off-the-top-of-the-head estimate of the answer.
6. Simplify the problem and do a simplified version first.

To make your first attempts at mathematical problem-solving easier, I will present each problem in three parts. In the "Problems" section that follows, I will just state the problem. If you get bogged down, there will be pointers in the "Hints" section starting on page 174. Whether or not you come up with a solution, you will be curious after a while to know the answer, which is given and fully explained in the "Solutions" section starting on page 175. *Good luck, and give yourself a fair chance!*

PROBLEMS

1. The head of a fish makes up one-fourth of the weight, the tail is one-eighth of the weight. The middle part weighs 20 pounds. Find the weight of the fish.

2. A $200 radio is subject to a 10% price increase by an over-confident merchant. The radio is not sold, and during the summer it is put on a 10% sale. What is the price of the radio during the summer sale?

3. In a purse there are nickels and dimes. (Doesn't this sound familiar?) There are twice as many dimes as nickels, and the total value of the coins is $5.25. How many nickels? How many dimes?

4. Add up all whole numbers from 1 to 100, as follows:

$$1 + 2 + 3 + 4 + 5 + \ldots + 97 + 98 + 99 + 100$$

Please try to find a method that avoids adding them up one by one. Even with a calculator, it is a lot of work and no fun.

5. A certain building project is expected to be completed in 25 days, given a crew of 8 workers. The deadline is unexpectedly moved up by 4 days, and 2 additional workers are hired. Will the deadline be met?

6. A radioactive substance has a half-life of 2 hours, which means that every 2 hours, half of the substance decays. What fraction of the original amount has decayed after 8 hours?

7. A merchant mixes 12 pounds of high-quality coffee ($6 per pound) with 8 pounds of standard coffee ($4.50 per pound). How expensive is the mixture?

8. The hero of this story is in a dark room and faced with the following task: he owns lots of socks, but only two kinds. He is partial to solid black and solid red socks.
   a. How many socks (minimum) does he have to take out of his sock drawer to be sure to come out with a pair of either black or red socks?
   b. How many socks (minimum) does he have to take out to be sure to wear black socks that day?

9. Eddy has two cats. Big, flabby Henrietta eats twice as much as elegant Emily. Together they polish off two cans of cat food a day (one can equals 6 ounces). How much does Henrietta eat?

10. An eccentric lady has a swimming pool with two faucets. The first faucet can fill the pool in 2 hours, while the second faucet takes 4 hours. One day the lady expects guests, and to speed up the procedure she turns on both faucets. How long does it take to fill the pool?

HINTS

1. Draw. Make the whole weight a circle. Draw the fractions.

2. Guess. If your guess is $200, it is as understandable as it is incorrect. Think again. Work out the price of the radio after the increase, and go step by step.

3. Imagine the coins. Most likely you don't like the "twice as many dimes as nickels" part. Don't close your eyes. Restate in your words. What does it really mean? It says that for every nickel there are two dimes. Imagine little piles of two dimes and one nickel each. Draw!

4. Look at the sum. $1 + 2 + 3 + 4 + \ldots 97 + 98 + 99 + 100$ for a while. Play with it. Can you group the numbers in any practical way? Keep trying. Simplify the problem. Try $1 + 2 + 3 + 4 + 5 + 6 + 7 + 8 + 9 + 10$ first.

5. My favorite way of doing "work problems" is simply to compute the number of man-days in the job. In this case that would be $8 \times 25 = 200$ man-days. There are 200 days of work to be considered. If one person was working alone, that is how long it would take him.

6. Compute the fraction of the substance that *remains* after 8 hours. Draw. Make the original amount a circle.

7. Estimate the mixture price. Will it be closer to $4.50 or $6? Then draw a concrete picture of the mixing bowl and what goes into it. How many pounds of coffee? How many dollars' worth?

8. a. Try different numbers of socks (2, 3, 4 . . .) and think about the possible color combinations.

    b. You are right. This problem cannot be solved unless I tell you how many socks of each color there are. Let us assume he owns 10 pairs of each color. Now you can do it!

9. Use brick arithmetic. Make Emily's portion the basic unit.

10. Instead of working with the filling times of the faucets, try to rewrite the problem in terms of speed. How *fast* does the first faucet work, etc. Speeds are additive, while there is no simple relationship between the different time spans. Take a guess for the time it will take.

SOLUTIONS

1. What you need to know to work out this problem is, what fraction (part) of the fish is the middle part?

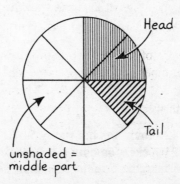

Head = $\frac{1}{4}$, tail = $\frac{1}{8}$; $\frac{1}{4}$ + $\frac{1}{8}$ = $\frac{3}{8}$. The middle part must be $\frac{5}{8}$. Now you know that $\frac{5}{8}$ of the weight amounts to 20 pounds. That means $\frac{1}{8}$ of the weight (1 slice of the pie) is 4 pounds. But then the total weight (the whole pie) must be eight times as much, or 8 × 4 pounds, or 32 pounds.

2. Original price $200
Increase 10%   +20
Increased price $220
Discount for
   summer sale 10%   −22
Sale price $198 (discount taken on increased
                    price)

Notice that the *percents* for increase and sale are the same, but that they are based on different prices, so they are different in dollar value.

3. The quickest way to solve this problem is to form little piles of 2 dimes and 1 nickel each. Each pile is worth a quarter (25 cents). How many piles can we make?

}   2 dimes and 1 nickel

pile

$25 \overline{)525}$ 21 piles = 21 nickels + 42 dimes

Check your answers:

42 dimes amount to $4.20
21 nickels amount to   1.05
                        $5.25

4. The most efficient way to arrive at this sum is the following:

$$1 + 2 + 3 + 4 + \bullet \ \bullet \ \bullet \quad + 97 + 98 + 99 + 100$$

Pair off the first and last numbers, the second and second last, etc. (see arrows). You will find that each pair adds up to 101. With 100 numbers, you will arrive at 50 pairs, each amounting to 101, that makes:

$$
\begin{array}{r}
101 \\
\times\ 50 \\
\hline
5{,}050
\end{array}
$$

5. In "Hints" we worked out that there are 200 man-days to the job. Now with 10 workers, 10 man-days are taken care of every day, which will make the job last for exactly 20 days. The time allowed for the job is 21 days, so there is no problem with the deadline.

6. After 2 hours there is ½ of the substance left.
After 4 hours there is ¼ of the substance left.
After 6 hours there is ⅛ of the substance left.
After 8 hours there is 1/16 of the substance left.

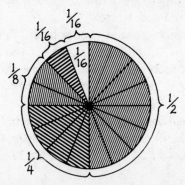

The original substance (the whole pie) = $^{16}/_{16}$.
So the part that decayed is $^{15}/_{16}$.

7. In the mixing bowl there are 20 pounds of coffee with a total value of $108 (12 × 6.00 + 8 × 4.50). The price of the mixture is 20)108, i.e., $5.40. Since there is more of the high-quality coffee, it makes sense that the mixture price is closer to $6. For an additional exercise, try 8 pounds at $4 and 12 pounds at $3.50. (Mixture price will come to $3.70.)

mixing bowl

8. a. He has to take out 3 socks. Obviously 2 socks are not enough. He might end up with a black-red combination. Every possible combination of 3 socks will include at least 2 of one color. He goes to work in black socks if he comes out with either 3 black socks or 2 black socks and 1 red; he goes to work in red socks if he has either 3 red ones or 2 red and 1 black.

   b. The given problem cannot be solved. There is not enough information. In "Hints," I suggested you assume he owns 10 pairs of each color. Now we know enough.

   To be sure he wears black socks, he has to think of the absolutely worst possibility in picking socks in the dark, namely that he might pick all the red ones first (20 socks). Then he needs 2 more, which will definitely be black. Answer: 22.

9. You most probably came up with the following drawing for Emily's and Henrietta's portions:

All three bricks together make 12 ounces. That means every brick is 4 ounces and Henrietta gobbles up 8 ounces a day.

10. Using the "Hint," we reformulate:
First faucet works at speed of ½ pool/hr.
Second faucet works at speed of ¼ pool/hr.

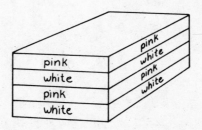

Together they work at (½ + ¼) pool/hr. = ¾ pool/hr. Now we have the speed. The question is, how much time? The answer is ⁴/₃ hours. Why? Imagine that our eccentric lady had her pool painted in 2 colors, each wall divided into 4 equal bands (see figure). If the faucets work at ¾ pool/hr. that means that after 1 hour the water covers 3 of these bands. There is 1 more band to go. How long did it take for 1 band: 20 minutes (3 bands in 60 minutes!), so the total time is 1 hour, 20 minutes (⁴/₃ hours).

Your persistence is commendable. You have worked through the basic chapters and you feel competent now to deal with the simple concepts you have met so far. The sure-fire way to learn the math for good is to go out and apply it. Practice comparison shopping, check discounts, and estimate the interest on your CD. The next part of the book, "Applying Math in Everyday Life," assists you in expanding your expertise. It is your opportunity to relearn and practice and it will make you streetwise in math.

V . . .
APPLYING MATH IN EVERYDAY LIFE

BE BOLD

You have learned most of the math you need to handle daily and commercial applications. The math involved in these situations is easy, involving addition/subtraction, taking percentages, or brick arithmetic.

Many questions that clients ask me have little to do with math and everything to do with commercial conventions. For instance:

What is a markup and how is it figured?
What is the Dow Jones Industrial Average?
How do I invest my money well?
What is the future value of my investment twenty years from
  now?

As a mathematician, I do not automatically know about banking, investing, or running a retail business. Knowing math enables me to figure the markup once the businessperson has given me the guidelines. I can figure future value of an investment once I'm given the predicted values of interest rate and inflation. I know little about the stock market, and I don't have that sixth sense for money and trends that make one a successful investor.

A friend recently asked me for help with pricing. Her job was selling wine to retailers. It turned out that she knew her math, but that her boss had neglected to give her complete information regarding the procedure. It was a purely procedural question. I can't walk into her wine company and do the pricing, but I would know what information I was lacking. I would have to go to my boss and say, "Here I have a shipment. Tell me exactly how this is figured from the vineyard to the store.'' That's not math. That's trade-specific or company-specific or store-specific practice. This example is a little like people asking me to do their income tax. I cannot do somebody's taxes just because I'm a mathematician. I need to study the law, the regulations. Neither the law nor the regulations are math. Executing the regulations and procedures is where the math comes in.

I am stressing this because knowing it gives you power. It makes math a smaller and easier subject. It frees you to ask how your company figures overhead, whether profits listed are net or gross, how your financial consultant ''knows'' that you'll be a millionaire by the year 2000.

Your company has guidelines and rules for overhead and markup—there is no general mathematical law. Your financial consultant makes *assumptions* about the future. *Nobody has secret knowledge that is beyond you.*

Approach the following situations lightly. Try out the math you have learned, or just watch it happen on the page. Draw, use brick arithmetic, estimate. Be bold.

1. FINANCIAL PLANNING (OR "THE 72 RULE")

Sabine, thirty-six, has $10,000 to invest. She is curious how much that money will be worth when she hits sixty. She is planning for a 24-year period. She has the money invested at 9%, but she has a somewhat riskier option at 12%. Use the "72 rule," the banker's shortcut to lengthy compounding.

THE 72 RULE

Divide the number 72 by the interest rate, and the result gives you the number of years it takes for your capital to double. Easy, isn't it?

In our case, we first check for 9%: $72 \div 9 = 8$; it takes 8 years for her money to double.

There are three 8-year periods in our planning range:

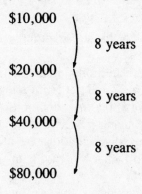

$10,000

        8 years

$20,000

        8 years

$40,000

        8 years

$80,000

Let's compare with the investment at 12%. To Sabine, this looks like a minor difference, just 3 pesky percents. Let's see what happens:

New interest rate: 12%. Using the 72 rule, we find that 72 ÷ 12 = 6: it takes 6 years for her money to double.

There are four 6-year periods in the 24-year planning range.

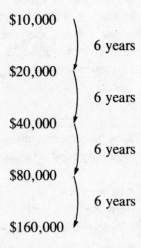

$10,000

6 years

$20,000

6 years

$40,000

6 years

$80,000

6 years

$160,000

Sabine is astounded that a 3% difference will make her investment twice as large.

Seasoned as we are in money matters, we know that Sabine won't be able to sit back and get rich on the wild growth of her original $10,000. We know that this looks too good to be true. How much will the buying power of that money be?

Let's assume an average inflation rate of 6%. (That's speculation, *not* math!) So Sabine's *actual* gains are at 9% − 6% = 3%. And 12% − 6% = 6%.

The actual growth rates for the two investments are 3% and 6% respectively.

3%

72 rule:
72 ÷ 3 = 24
(money doubles
every 24 years)

$10,000
) 24 years
$20,000

6%

72 rule:
72 ÷ 6 = 12
(money doubles
every 12 years)

$10,000
) 12 years
$20,000
) 12 years
$40,000

So the actual buying power, in 1990 dollars, is $20,000 and $40,000 respectively. The decision about which investment to choose has obviously nonmathematical aspects. It is likely that the 12% investment has more risks attached, and there are tax questions to be considered as well as other concerns that are in the realm of accountants, financial planners, and investment counselors.

Math helps you figure out the different options, given the assumptions like future interest rates, inflation, and stock market trends. It cannot make the choices for you.

The sixth sense some mathematicians have for numbers and patterns is *not* the same as the sense of money and good business opportunities. I have met many investors and bankers who have an unfailing "nose" for trends and opportunities and at the same time have average or below-average math skills.

2. ANNUAL RATE AND ANNUAL YIELD

Above is an ad with a scribbled question that I found one day
on my desk from my friend Joe. He was frustrated that he did
not understand the numbers *or* the explanation!

When I was a child in Switzerland, banks were simpler and
so was interest. Come December 31, everybody would undertake the pilgrimage to the bank to have the interest earned during
the year entered into his bankbook and added to the capital.

These days, most banks compound daily. This means that every day Joe's interest is figured at 8.25% and added to his capital. So the next day the increment will be slightly larger, since it's taken on a larger base. If Joe leaves his money for a whole year, his capital creeps up slowly, and so do the daily increments. This combined movement results in a higher return on his money. In Joe's case, that return is 8.60% and it's called *annual yield*.

Let's see how much of a difference this makes in a year on Joe's $10,000 savings (use a calculator!):

1. *No compounding:* 8.25% on $10,000
Keying sequence: 10000 ⨉ 8.25 %
Joe gets $825 interest.

2. *Daily compounding:* 8.60% on $10,000
10000 ⨉ 8.60 %
Joe gets $860 interest.

Hint: Ads often make it sound as if your bank is making you a present by compounding daily. Of course, the bank first figures what annual yield it is willing to pay, and fixes the rate after that.

3. PAY INCREASE

Carol is making $20,000 annually as an administrative assistant, and she has worked extremely hard all year. She marches into her salary review, determined to ask for a $1,500 increase. After she is finished, her boss makes her a counteroffer:

"I'll give you a bonus of $700 right now and a 4% increase." Carol's head is spinning. What percent increase was *she* asking for? Was the boss's offer better or worse? What should she say?

We all know what she should say: "Thank you, I'll think about it and get back to you tomorrow." (Much better, of course, to have thought of the possible problems beforehand.)

1. What percent raise is Carol asking for?
   Use the 1% method:
   1% of $20,000 is $200
   $1,500 is how many times $200?
   $1,500 \div 200 = 7.5$
   She is asking for 7.5%.

2. How good is the boss's offer?
   Raise: 4% of $20,000
   1% of $20,000 is $200
   4% of $20,000 is 4 × $200, or $800
   $800 + $700 bonus = $1,500

So the cash value of both options is the same. If Carol wants some cash fast, she'll accept the boss's offer. If she thinks of the future, she'll insist on her demand because a higher raise results in a higher salary next year, and her next raise will be figured on that larger base.

▪ ▪ ▪
4. SUCCESSIVE DISCOUNTS

Harry works at a furniture store. One Saturday he brings in his family to choose a new sofa. After much debating, they settle on a three-seat black leather sofa originally priced at $2,000. Today it's on sale at 30% off. Harry gets a 10% employee discount, and there is also a 2% cash discount.

"Wow," little Chip says. "We'll get the sofa 42% off. That's almost half-price."

Is Chip right?

Let's think about it. Walk through it. First you choose the sofa. Then the people at the store take off the 30%. Then they'll smile at Harry and take off his 10%. Then he tells them he'll pay cash, and they take off 2%.

Let's do it:

Price	$2,000.00
Discount (30%)	− 600.00
Sale price	$1,400.00
Employee discount (10%)	− 140.00
Payment due	$1,260.00
Cash discount (2%)	− 25.20
	$1,234.80

Notice that every discount is taken on a different and successively lower base, so Chip was wrong. The discount is less than 42%. In case you worried about which discount to take first or last, the good news is that the sequence does not matter.

Experiment with $100; it is always a good idea to analyze general percent problems.

Chip thinks that a 30% discount and a 10% discount amount

to a 40% discount. We realize now that he is wrong, but what percent is it really?

$$
\begin{array}{r}
\$100 \\
-\ 30 \\
\hline
70 \\
-\ \ 7 \\
\hline
\$\ 63
\end{array}
$$

That amounts to a discount of $37 on $100. So successive discounts of 30% and 10% amount to a combined discount of 37%.

5. PRORATE

Brenda has recently been made a supervisor, and now she has to figure benefits for her employees. Carlo is leaving on June 1, and his benefits are 19 vacation days and a $3,000 bonus.

The company prorates vacation days and bonus. (*Prorate* means you get benefits proportionately to the part of the year you spent with the company; i.e., for six months you get half the benefits.)

Let's do the bonus first.

Carlo leaves on June 1. That means he has worked 5 months out of the full 12 months of the year. His bonus for the full year is $3,000.

Let's do proportions the European way. Go to the basic unit of 1 month:

For 12 months, Carlo gets $3,000 bonus.

For 1 month, he gets $3,000 ÷ 12 = $250.

For 5 months he gets 5 × $250 = $1,250.

We'll use the same procedure for the vacation days, but since 19 days is not nice and round, we'll use the calculator. It's the same procedure: divide by 12 for one month, and then multiply by 5 for 5 months.

Here is the keying sequence:

$$19 \boxed{\div} 12 \boxed{\times} 5 \boxed{=} 7.9 \text{ vacation days}$$

So Carlo gets 8 vacation days and a $1,250 bonus.

■ ■ ■
6. FOREIGN CURRENCY

Janet is planning a trip to France. Her clothes are ready, the
house-sitter is in place. All's well, except that Janet forgot to get
French francs. She wants to be clear this time how much French
money is worth. She remembers her last trip, when she spent
French francs like Monopoly money, and took months to re-
cover from the resulting Visa bill.

She picks up the business section of her daily paper and looks
up "Foreign Exchange."

Under French francs, it says:

	FGN CURRENCY IN DOLLARS	DOLLAR IN FGN CURRENCY
France (franc)	.1520	6.5800

How can Janet quickly figure how much the bank should give
her in exchange for $200?

How can Janet assess how good a deal her bed-and-breakfast
in Paris is, at 400 francs per night?

From the second column in the newspaper, Janet sees that $1
is about 6 francs.

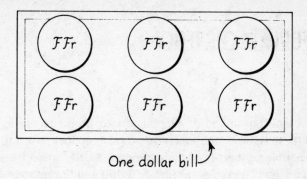

One dollar bill

So her $200 will be approximately 200 × 6 francs = 1,200 francs.

Going the other way, 1 franc is about 15 cents (first column). So the bed-and-breakfast in Paris is about 400 × $.15 = $60.

Janet is pleased with herself for having found out how to answer her own questions about French currency. She promises herself to visualize those six francs sitting on the dollar bill to keep her spending in check.

7. BUDGET

George and Susan had developed serious problems around their budget. Along with the marriage vows, they had sworn to share all expenses of their household *proportionately* to their income. At this point, George is earning $50,000 and Susan $30,000 per year. He insists that they share every expense 60%/40%. Susan is almost sure that at 40% she is overpaying her share, but she does not know how to figure the correct percent figure. Is she right?

Very few people go to the trouble these young people are putting themselves through, but it is their idea of what is equitable, so let's solve their problem:

George is making $50,000, Susan is making $30,000, so for every $5 of George's, Susan makes $3.

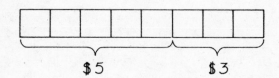

Look at the picture; it tells you that out of every $8 of family income, George contributes $5 and Susan $3. So Susan has to pay $3 out of every $8. That is ⅜, which amounts to 37½% (¼ = 25%; ⅛ = 12½%; ⅜ = 37½%).

So Susan is right! She has been overpaying at her previous 40% level.

Now let's look at their budget.

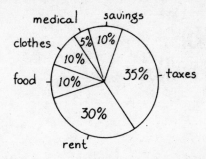

The circle graph is an easy way to show a budget. You can think of the whole circle either as $1 (often used by the government; that's how the federal tax dollar is spent) or, as I prefer for this case, as symbolizing the total income.

The slice for rent is 30%. Let's figure their rent in dollars: 10% of $80,000 is $8,000, 30% of $80,000 is *$24,000.*

Now you figure their major expenses:

Food: _____
Clothes: _____
Medical: _____
Taxes: _____

If their rent increased by 20%, what percent slice would "rent" occupy after the increase?

So far, rent is 30%. Increase by 20%.

If 10% of 30% is 3%, 20% of 30% is 6%.

The *new* rent slice would be 36%.

Answers for expenses, above:

Food: $ 8,000
Clothes: $ 8,000
Medical: $ 4,000
Taxes: $28,000

The sum of $8,000 is left for miscellaneous expenses and savings.

8. HOW MANY
AT THE WEDDING?

Math shows up in unexpected places. This lovely word problem I owe to television and radio talk-show host Sally Jessy Raphael. Listening to the radio at two o'clock one morning, I heard a caller tell Sally her problem. She had just mailed 120 invitations for her wedding. How many guests should she actually expect? Is there a general rule about how many people accept invitations? Ticking off numbers and percentages, Sally—at that hour!—gave the rule-of-thumb answer:

Weddings:	75% attendance
Parties:	50% attendance
Dinner in Manhattan:	25% attendance

Sally went on: "If you send 120 invitations, around 90 guests will show, and if you want 120 guests, you need to mail 40 extra invitations."

How did Sally do it in her head? I don't know, but here's how you could do it in your head.

Remember those famous pies?

QUESTION 1:

How many guests can be expected if 120 invitations are sent?

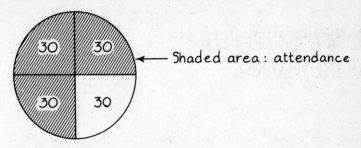

The whole pie represents the 120 invitations.

Answer: 90 people will show up.

QUESTION 2:

Attendance, 120 guests. How many invitations?

In this question, the 120 is not the whole pie. We don't know the total number of invitations.

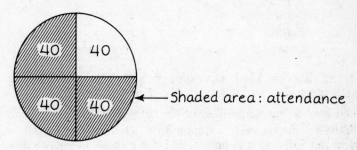

The whole pie represents total number of invitations.

Answer: 160 invitations are a good bet.

With my kind of luck, if I were sending the invitations, everyone would show up and there wouldn't be enough champagne!

9. HOUSEHOLD HELP

Your full-time baby-sitter or housekeeper charges you $6 per hour. If your overall tax rate is 25%, how much do you actually have to earn to cover the cost of your help?

Let's assume the baby-sitter comes for 40 hours per week; that is, she gets 40 × $6 = $240 per week. Let's draw circles:

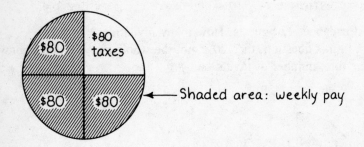

Three slices = $240. One slice = $80. Total circle is 4 × $80 = $320.

You have to earmark $320 for the baby-sitter.

What if your tax rate is around 35%? (Estimate at 33⅓% = ⅓.)

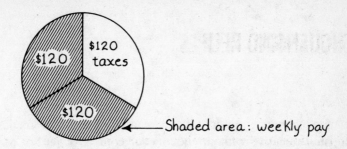

Shaded area: weekly pay

Two slices make $240. One slice is $120. Total circle is 3 × $120 = $360.

This method is obviously helpful for estimating whether you can afford private school or college for your child and, in fact, in every case where your expenses are not tax-deductible.

10. CHANGING RECIPES

Amanda had enough to do in her demanding job as a cosmetics-company executive without cooking for company. Yet her husband Steve thought it would be a good idea to invite two couples. So here was Amanda, standing in the kitchen with a recipe from *The New York Times Magazine*, "Chicken Breasts with Garlic and Balsamic Vinegar," that she would like to try if ever she had time. The recipe served four, and she would have to increase it to serve two more. How? Should she double it and freeze the leftovers, or juggle the recipe so it would serve six?

Here's the recipe as printed:

CHICKEN BREASTS WITH
GARLIC AND BALSAMIC VINEGAR*

4   skinless boneless chicken breasts, halved, about
     1 1/4 pounds in all
Salt and freshly ground pepper to taste
3/4   pound small- to medium-sized mushrooms
2   tablespoons flour
2   tablespoons olive oil
6   cloves garlic, peeled
1/4   cup balsamic vinegar
3/4   cup fresh or canned chicken broth
1   bay leaf
1/2   teaspoon minced fresh thyme or 1/4   teaspoon dried
1   tablespoon butter

*Copyright © 1989 by The New York Times Company. Reprinted by permission.

1. If the chicken breasts are connected, separate the fillets and cut away any membranes or fat. Sprinkle with salt and pepper.
2. Rinse the mushrooms, drain and pat dry.
3. Season the flour with salt and pepper and dredge the chicken breasts in the mixture. Shake off the excess.
4. Heat the oil in a heavy skillet over medium-high heat and cook the chicken breasts until nicely browned on one side, about 3 minutes. Add the garlic cloves.
5. Turn the chicken pieces and scatter the mushrooms over them. Continue cooking, shaking the skillet and redistributing the mushrooms so that they cook evenly. Cook about 3 minutes and add the vinegar, broth, bay leaf and thyme.
6. Cover tightly and cook over medium-low heat for 10 minutes. Turn the pieces occasionally as they cook.
7. Transfer the chicken to a warm serving platter and cover with foil. Cook the sauce with the mushrooms, uncovered, over medium-high heat for about 7 minutes. Swirl in the butter.
8. Discard the bay leaf. Pour the mushrooms and sauce over the chicken and serve.

Yield: Four servings

Here's how Amanda changed the recipe to serve eight and to serve six. The fractions in the recipe scared Amanda, and she figured that doubling (to serve eight) would be easier, so she wrote out the three fractions that show up in this recipe: $\frac{1}{2}$, $\frac{1}{4}$, $\frac{3}{4}$.

That wasn't so bad.

She tried to double each one: double $\frac{1}{2}$ is 1; double $\frac{1}{4}$ is $\frac{1}{2}$.

Then she came to $\frac{3}{4}$. "Double $\frac{3}{4}$ is ?" is what she wrote.

Suddenly Amanda thought "money": 3 quarters and 3 quarters = 6 quarters = \$1.50 = 1½ dollars. So, double ¾ is 1½.

Done. Whew.

Emboldened by her success, Amanda decided to cook for six and be done with it. How could she go from 4 to 6?

She doodled:

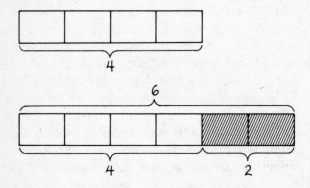

So cooking for 6 was like cooking for 4 and for 2, one and a half times the original recipe. She worked out the whole numbers first:

	FOR 4	FOR 2	FOR 6
Chicken	4	2	6
Flour	2	1	3
Oil	2	1	3
Garlic	6	3	9
Butter	1	½	1½
Fractions			
Thyme	½	¼	¾ (not bad!)
Vinegar	¼	⅛	⅜
Mushrooms	¾	?	?
Broth	¾	?	?

Amanda liked her list, and she decided to find the missing quantities for mushrooms and broth by drawing the fractions out in pie form:

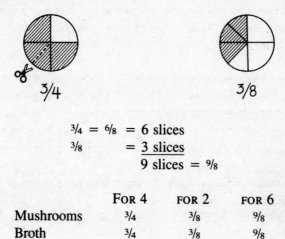

3/4   3/8

$$3/4 = 6/8 = 6 \text{ slices}$$
$$3/8 \quad\quad = \underline{3 \text{ slices}}$$
$$9 \text{ slices} = 9/8$$

	FOR 4	FOR 2	FOR 6
Mushrooms	3/4	3/8	9/8
Broth	3/4	3/8	9/8

Needless to say, Amanda was proud of herself. The doodling, the bricks, and the circles helped. It made the numbers look reasonable and within her grasp.

11. AREA AND VOLUME

John has just moved to New York City, and rented a large studio in SoHo. He wants to get settled quickly. He needs carpeting, and since he comes from Minnesota, he knows he'll need an air conditioner first thing. At the department store he sees an air-conditioning unit that promises to be energy-efficient and low-noise.

"How many BTUs do you need?" asks the salesman.

"I don't know," John answers.

"Just tell me the volume of the room you want to cool, how many cubic feet?"

John quietly withdraws. He does not know BTUs, and certainly not the volume of his studio. He just arrived. *Gimme a break*. He buys a tape measure and goes home. An architect friend explains on the phone, "Just measure the length, width, and height of your studio, and multiply them. Easy."

John's studio length, 21 feet; width, 15 feet; height, 10 feet.

$$\text{Volume} = 21 \text{ feet} \times 15 \text{ feet} \times 10 \text{ feet} = 3,150 \text{ cubic feet}$$

From *Consumer Reports* John gets the recommended number of BTUs and finds out, by the way, that the highly touted unit at the department store ranked low in the magazine's ratings of energy efficiency.

Now for the carpeting. John vaguely remembers from school that the area of his floor (an actual rectangle, with no nooks or crannies) could simply be figured by multiplying length times width:

Area = 21 feet × 15 feet = 315 square feet

Armed with all that information, John takes another trip, buys his air conditioner, and finds the carpet he wants at $22 per square yard. He figures 1 square yard is 3 square feet. That means he needs 105 square yards, and the carpet will cost him a fortune. Just then a saleswoman comes to his rescue. She corrects his misconception and assures him that 1 square yard is 9 square feet, so his apartment, at 315 square feet, is 35 square yards.

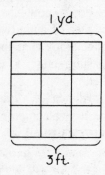

1 square yard = 9 square feet

That brings the price down to a more reasonable 35 × $22 = $770.

12. BALANCING YOUR CHECKBOOK

Balancing your checkbook sounds like an act of penance. People feel guilty if they don't do it and stupid if they don't know how. It seems that bank statements awaken all our fear of authority and all our math anxiety. Yet balancing your checkbook is mathematically easy. All it involves is adding and subtracting. It is work, yes, it is tedious, yes, but it is not difficult.

Let's think for a moment. What is there to balance? Why do we have to figure anything? What does the bank have a computer for? What are the reasons the balance in your checkbook and the one on your bank statement don't agree?

First, there are things the bank does not know about you, such as deposits in transit and cash withdrawals after date of statement. Second, there are things you don't know about the bank, such as outstanding checks and service charges.

So balancing your checkbook is really much more like a detective story than a math problem. The bank sends you a form with complete instructions. Basically, you list and add the outstanding checks, and list and add what you spent after the closing date of the statement. Then you add or subtract these totals according to the instructions.

You got it.

If, after careful evaluation, the two balances don't agree, either you or the bank made a mistake, more likely you. Check your additions with a calculator. Notice if you reversed any digits on your checks or stubs.

Hint: If the disparity is small, do yourself a favor and let it go.

If the discrepancy is *substantial*, however, the amount usually gives you a clue as to what is wrong. I once withdrew $200 from an account and immediately changed it into Finn marks. My bank statement was off by exactly $200, so the mistake could be easily traced. The bank had subtracted my $200 twice, once in dollars and once in Finn marks.

13. SAMPLE MATH TEST
FOR EMPLOYMENT

My friend Gilbert needed some extra cash over the holidays and decided to do the easiest thing: get a department store cashier's job. Gilbert was a respected researcher for a TV station, and was sure that he would get the job. To his considerable surprise, he got a test slapped in front of him, a math test that looked difficult to him. Here is a copy.

1. What is the total cost of two pencils at $.10 each and one dozen erasers at $.40 a dozen?

2. If peaches are selling at $.40 a can or $4.60 per dozen cans, how much is saved on each can by purchasing the dozen cans?

3. One room in an office has 9 rows of filing cabinets with 9 filing cabinets in each row. Another room has 6 rows of cabinets with 12 cabinets in each row. How many more cabinets are there in the first room?

4. A detergent sells at $.75 a quart. How much will 5 gallons cost?

5. The average cost per pound of bananas, grapes, and plums is $.30. If bananas cost $.10 a pound and grapes cost $.45, how much do plums cost per pound?

6. How many packs of envelopes can be bought for $3 at the rate of 2 for $.50?

7. At 8:00 A.M. the barometric pressure was 30.6 and at 11:00 A.M. the pressure was 31.8. Assuming a constant rate of increase, what time was it when the pressure was 31.0?

8. A man earned $28 and saved $7. What percent of his earnings did he save?

9. The premium for $1,000 of insurance is $50. What is the premium for $5,500 of insurance?

10. A man worked one week from 9:00 to 4:00 with 30 minutes for lunch, and Saturday from 9:00 to 1:00. How many hours did he work that week?

11. One dealer offers a 25% discount on a $100 filing cabinet. Another offers successive discounts of 20% and 10% on the same filing cabinet. What is the difference between the net prices?

12. A man earns $20 for producing 40 units of work. For each unit produced over these 40, he is paid at the rate of 1½ times his regular rate. How much will he receive if he produces 52 units of work?

13. If a self-employed writer earns $2,400 and he can deduct $1,500 for office rent and $200 for other expenses, how much state tax does he pay if the balance is taxed at the rate of 5%?

14. A student gets ⅔ of a cent for each leaflet he hands out, and he gets rate and a half for every one over 2,400. How much does he earn for handing out 3,100?

15. At the end of his sixth year with a company, an employee received a bonus of $2,500, which was twice as much as he had received at the end of his first year. If his bonus in-

creased an equal amount each year, what was his bonus at the end of the fourth year?

Answers: (1) $.60; (2) 1⅔ cents; (3) 9; (4) $15; (5) $.35; (6) 12; (7) 9:00 A.M.; (8) 25%; (9) $275; (10) 36½; (11) $3; (12) $29; (13) $35; (14) $23; (15) $2,000.

CONCLUSION

If you have worked your way through the thirteen situations in Part V, you realize how much you have learned, and now, having your basic math in order, you can be assertive in new situations, however intimidating they might have been to you before.

Let me tell you about something that happened to a couple I know. A real-estate salesperson tried to sell them a particular property, insisting that the rent they would earn from the property would easily cover their mortgage and maintenance. When my friends expressed doubts, the salesperson covered four pages of paper with numbers to prove it. If this happens to you, STOP. Breathe. Be smart and assertive. Ask for the papers and go home and check them with a trusted soul (preferably a banker or a CPA). That's what they did because they knew that the salesperson's insistent repetition of "Don't you see? Don't you see?" was meant to intimidate them and hurry them along. There are many ways to figure the feasibility of an investment, and salespeople tend naturally to take an optimistic view of things.

The only kind of math in which you absolutely need to be competent as a regular citizen and consumer is the math explained in Part IV. If you want to go beyond this basic math, it would pay to take a course or read a good self-study book on business math. (See the list of "Recommended Reading" on pages 221–22.)

VI...
AFTERTHOUGHTS

7.. THE RIPPLE EFFECT

Whether you've just leafed through this book or actually worked through every problem, you have taken steps to overcome your math anxiety. You may already feel a little different when your friends mention a 5% pay raise or that the Dow Jones has gone up by 3 points, or that the weatherman predicts a 70% probability of rain. You no longer need to run and hide from those pesky numbers.

You may also notice that your former apprehension around numerical information has given way to a new assertiveness. You expect to understand math now, and you take all the steps it takes to get there: you listen, you read, you read again, you ask questions. You know that your questions are relevant and worth answering.

Along with the assertiveness comes a helpful irreverence toward those numbers that are used to manipulate us: inflated or skewed statistics as they are used in politics and advertising; the various numbers that people use to label us, such as grades, IQ, and medical risk factors.

You question whether it's really true that 35 million Americans have serious arthritis or that 25% of the population suffers from migraines. You don't just believe the numbers the experts and the media hand you. You go and check them.

Once you can bandy about numbers and receive them with a healthy skepticism, you will find that you notice a pleasant "ripple effect'' into other areas of life. People who suffer from math anxiety often carry a hidden agenda through life. Deep down, they give magical power to math and those who know it.

As a consequence, math anxiety very often infiltrates areas of function that to our unconscious mind seem like math. As long as you were afraid of math, you may have developed a fear of mastery, of detail, of being wrong, of overstepping boundaries, of invading male territory (if you are a woman), of giving yourself power, of growing up, of dealing with money, of being boring, of facing Judgment Day, of being found a fraud.

You'll probably identify with more than one of the above fears. Since for you they were connected with math anxiety, the ripple effect will now sweep them away. See if, in the next few months, you don't experience a new freedom born of your competence and confidence with numbers. You might find it easier now to speak up at meetings, to get your finances in order, and to take personal or business risks, because going out on a limb is now just one more adventure.

Vanessa, whose journey through Mathland we followed from her very anxious beginnings, went on to the graduate school of her choice and mastered an intensive three-semester statistics requirement. She told me that the Confidence-Building Techniques (particularly the ''panic stuff'') were as useful as her newfound ability to argue, draw, ask, debate, and read her way through any math that came along. Having mastered more math than most mortals ever see or need, Vanessa is experiencing the ripple effect as a strong ego boost: the knowledge that she can handle just about anything, including her finances, her independence from her family, and her considerable academic success.

I hope this book will be useful to you beyond learning basic math. The way we look at learning in this book, the Confidence-Building Techniques, and the problem-solving strategies are applicable in any numerically challenging situation. Come back to the book when you need a confidence boost or for a simple review of math. Pick it up when you face new and bigger math challenges.

What I've tried to do is to teach you to cheer yourself on through the difficulties of overcoming your math blocks, and actually deal with the math that used to frighten and annoy you. I've tried to give you tools that go far beyond this immediate task, tools for lifelong learning. They are:

- Analyzing the actual problem.
- Discovering your feelings and blocks.
- Building up your confidence through specific techniques that let you proceed in spite of fear and trepidation.
- Adopting a down-to-earth approach to new material, translating jargon into everyday language, using drawings and props to get a better look at problems.
- Developing problem-solving strategies useful far beyond the subject matter of this book.

It is important to remember that learning is not an activity restricted to children and students. In our society, rapid technological and social changes demand that we keep mentally fit and open to new information. This is particularly true of math. The vast information flow (or overflow!) has to be treated with intelligence. We cannot and do not want to remember every number and every statistic we meet. We *do,* however, want to sort out what's important and what isn't, what is math and what isn't, what we need to retain and what we need to let go.

Just last week I ran into a former client on a subway platform. She was waving a computer printout at me.

"You won't believe this," she began. "Yesterday afternoon I could not really make sense of these numbers, so I decided to take them home and look them over. I fell asleep over them last night, unable to solve my problem. Early this morning I did my favorite affirmation, 'I am brilliant.' While waiting for the train this morning, I pulled out the printout. It looked new and clear to me, no trouble at all.'' She went on, ''It really works. I just kept thinking and kept up my confidence, and now I'm ready for my day at work.''

■ ■ ■
RECOMMENDED READING

ON MATH AND RELATED ANXIETIES

Friedman, Martha. *Overcoming the Fear of Success.* New York: Warner, 1980.

Jeffers, Susan. *Feel the Fear and Do It Anyway.* New York: Fawcett, 1987.

Kogelman, Stanley, and Joseph Warren. *Mind over Math.* New York: McGraw-Hill, 1978.

Miller, Alice. *The Drama of the Gifted Child.* New York: Basic, 1983.

Tobias, Sheila. *Overcoming Math Anxiety.* Boston: Houghton Mifflin, 1980.

Wurman, Richard Saul. *Information Anxiety.* Garden City, N.Y.: Doubleday, 1989.

THE BRAIN, LEARNING, AND CONSCIOUSNESS

Bandler, Richard. *Using Your Brain for a Change.* Berkeley: Real People Press, 1985.

Belenky, Mary Field, Blythe McVicker Clinchy, Nancy Rule Goldberger, and Jill Mattuck Tarule. *Women's Ways of Knowing: The Development of Self, Voice and Mind.* New York: Basic, 1986.

Brown, Barbara. *Supermind.* New York: Harper & Row, 1979.

Buzan, Tony. *Use Both Sides of Your Brain.* New York: Dutton, 1974.

Edwards, Betty. *Drawing on the Right Side of the Brain.* Los Angeles: Tarcher, 1983.

Feldenkrais, Moshe. *Awareness Through Movement.* New York: Penguin, 1980.

Gawain, Shakti. *Creative Visualization.* New York: Bantam, 1982.

Goldberg, Philip. *The Intuitive Edge: Understanding and Developing Intuition.* Los Angeles: Tarcher, 1983.

Hampden-Turner, Charles. *Maps of the Mind*. New York: Macmillan, 1981.

Krishnamurti. *On Education*. New York: Harper & Row, 1974.

Ornstein, Robert. *The Psychology of Consciousness*. New York: Penguin, 1972.

Schroeder, Lynn, and Sheila and Nancy Ostrander. *Superlearning*. New York: Delta, 1979.

Shah, Idries. *The Pleasantries of the Incredible Mullah Nasrudin*. New York: Dutton, 1983.

Suzuki, Shunryu. *Zen Mind, Beginner's Mind*. New York: Weatherhill, 1970.

IF YOU WANT TO GO ON WITH MATH

Butts, Thomas. *Problem Solving in Mathematics*. Glenville, Ill.: Scott Foresman, 1973.

Carman, Robert A., and Marilyn J. Carman. *Quick Arithmetic*. New York: John Wiley, 1984.

Huff, Darrell. *How to Lie with Statistics*. New York: Norton, 1954.

Meyer, Robert J. *Consumer & Business Math*. New York: Arco, 1974.

Polya, George. *How to Solve It*. Princeton, N.J.: Princeton University Press, 1945.

MATH IN CONTEMPORARY AMERICA

Everyone Counts: A Report to the Nation on the Future of Mathematics Education. Washington, D.C.: National Research Council, 1989.

Paulos, John Allen. *Innumeracy: Mathematical Illiteracy and Its Consequences*. New York: Hill & Wang, 1988.

MATH TO PONDER

Doczi, Gyorgy. *The Power of Limits: Proportional Harmonies in Nature, Art and Architecture*. Boulder and London: Shambhala, 1987.

Gleick, James. *Chaos*. New York: Penguin, 1987.

Hofstadter, Douglas R. *Gödel, Escher, Bach: An Eternal Golden Braid*. New York: Vintage, 1979.

INDEX

Mind-dump exercise, 73–74
Minorities
  math anxiety and, 9–11
  sideways thinking and, 51–52
Moral failing, math deficiency as, 47
Multiplication
  decimals, 116–17
  fractions, 127–28
  tricks for, 90–91
  zero-ending numbers, 121
Murdoch, Rupert, 15
Music, 71–72
Myths about math, 41–53

National Research Council, 7
Negative criticism, self-image and, 64
Number sense, 118
Number system, 97–98

1% method, 109–10
One-page-a-day math, 93–94
Open-ended word problems, 160–62, 165–72
Origami, 151–57
Ostrander, Nancy, 71–72
Ostrander, Sheila, 71–72

Panic Kit tools, 63–69, 75–79
Passivity, as math block, 56
Patterns, 88–90
Paulos, John Allen, 8–9, 17, 57
Pay increase, 190
Percents, 104
  conversions, 146–48
  1% method, 109–10
  round percentages, 107

savings account, 111–12
  10% method, 106–10
  tipping, 104–5
Perfectionism, as math block, 55
Perimeter, area and, 155–57
Problems
  break-taking in solving of, 51
  chain problems, 95–97, 104, 114, 122–23, 132, 141, 144, 151, 159
  formulas and, 42–43
  magnitude problem, 119
  word problems. *See* Word problems
Proportional leaps, 142–43
Proportions, 141–42
Prorate, 193
Puzzles, 100–102, 128–30

Quiz
  for evaluation of math anxiety, 25–32
  math, 85–86

Reading, of word problems, 163
Recipes, changes in, 202–5
Regression, as math block, 55–56
Relational learning style, 49
Rewards, 72–73
Rewriting, of word problems, 164
Roman numerals, 98
Rote learning, 21
Round percentages, 107

Savings account, 111–12
Schroeder, Lynn, 71–72
Selective math anxiety, 19–20